Catalunya, One Nation, Two States

Catalunya, One Nation, Two States

An Ethnographic Study of Nonviolent Resistance to Assimilation

Alexander Alland, Jr.
with
Sonia Alland

CATALUNYA, ONE NATION, TWO STATES
© Alexander Alland, Jr. with Sonia Alland, 2006.

All rights reserved. No part of this book may be used or reproduced in any manner whatsoever without written permission except in the case of brief quotations embodied in critical articles or reviews.

First published in 2006 by
PALGRAVE MACMILLAN™
175 Fifth Avenue, New York, N.Y. 10010 and
Houndmills, Basingstoke, Hampshire, England RG21 6XS
Companies and representatives throughout the world.

PALGRAVE MACMILLAN is the global academic imprint of the Palgrave Macmillan division of St. Martin's Press, LLC and of Palgrave Macmillan Ltd. Macmillan® is a registered trademark in the United States, United Kingdom and other countries. Palgrave is a registered trademark in the European Union and other countries.

ISBN-13: 978–1–4039–7439–6 (hardcover)

ISBN-13: 978–1–4039–7440–2 (paperback)

Library of Congress Cataloging-in-Publication Data

Alland, Alexander, Jr. 1931–
    Catalunya, one nation, two states : an ethnographic study of nonviolent resistance to assimilation / Alexander Alland, Jr. with Sonia Alland.
      p. cm.
    Includes bibliographical references and index.
    ISBN 1–4039–7439–X (hc: alk. paper)—ISBN 1–4039–7440–3 (pb: alk. paper)
    1. Nationalism—Spain—Catalonia. 2. Catalans—Ethnic identity. 3. Catalans—Cultural assimilation—Spain—Portbou—Case studies. 4. Catalans—Cultural assimilation—France—Cerbère—Case studies.
    I. Alland, Sonia. II. Title.
DP302.C69A55 2006
320.540946′7—dc22   2006043266

A catalogue record for this book is available from the British Library.

Design by Newgen Imaging Systems (P) Ltd., Chennai, India.

First edition: December 2006

10 9 8 7 6 5 4 3 2 1

Transferred to Digital Printing in 2011

Contents

Preface		vii
Introduction		1
Chapter One	The How and Why of This Book	7
Chapter Two	Framing the Study: The Origin and Meaning of Nationalism	37
Chapter Three	The People of Cerbère Speak	65
Chapter Four	The People of Portbou Speak	93
Chapter Five	A New Direction	115
Chapter Six	The State of Catalan in Rosselló (Roussillon)	133
Chapter Seven	Northern Catalans Speak	149
Chapter Eight	Language and Identity on the Horns of a Dilemma	175
Appendix		191
Bibliography		197
Index		203

Preface

This is a book about Catalunya, a nation without a state encompassing a past that goes back at least to the tenth century. It is also about the nonviolent struggle of Catalans to maintain their culture and language in the face of two powerful nation-states, France and Spain.

Our research began in a more or less traditional anthropological setting, which is to say, in two small populations of approximately 1,500 each. The size of these villages made it possible to use the principal method of anthropological research, namely participant observation. This method allows patient researchers to at least partially integrate themselves into the host culture through the establishment of social relations with informants.

Apart from reasons that are explained in the body of this work we chose these particular villages, Portbou and Cerbère, not only for their size but also because both are Catalan in origin and are located about a mile apart on the frontier between France and Spain. Although both communities are *historically* Catalan, only Portbou is part of the Catalan autonomous region of Spain, the *Principat*, whereas Cerbère is now, in every sense of the term, an integral part of the highly centralized French state. Before we discovered these facts doing a comparative study of these two villages looked particularly promising. Both were founded in roughly the same year at the end of the nineteenth century, and for the same reason, as rail-heads to solve the inconvenience of different track widths in the two countries. This difference required the transboarding of products from one rail system to the other. Consequently, Cerbère and Portbou rapidly became prosperous. In addition to the manual labor required for the transfer of merchandise, it was easy for the growing population to find work in the custom services or the many private businesses that quickly sprang up due to the heavy burden of customs regulations. In addition, as each village lies at the foot of rugged mountains making it difficult to catch smugglers, a large contingent of national police on both sides of the border was necessary to enforce the rules of commerce. These facts seemed to present us with an ideal controlled field project comparing Catalan culture in two different nation-states. Furthermore, we were aware that an economic crisis had recently struck both villages, when the European Community abolished border controls across the frontiers of several member states including France and Spain. As a result, unemployment increased in both villages causing a rapid loss of about half their populations in a few years time. As a response, the mayors of Cerbère and Portbou, both Catalan speakers, met to develop a set of projects linked to a hoped-for resource—that is, increased tourism based on the exploitation of a common Catalan culture.

As it turned out we were quite naive about the possibilities of cooperation between the villages in the face of state interest, particularly in France. Also we did not realize the extent to which the French government had diluted Catalan culture on its side of the border. Once in place we discovered that although Portbou had retained its cultural status as Catalan with strong links to the *Principat* (the historical center of Catalan culture with its capital in Barcelona), the French had done a rather thorough job of folding Cerbère into the French state.

Nonetheless, still wishing to investigate the state of Catalan culture in both countries, we widened our investigation to include a more general study of Catalan identity in the *Principat* as well as the "Catalan" region of France, the department of the Roussillon.

As a result, what was initially to be a rather typical locally based study became an *untypically* large undertaking for anthropologists. In what follows we explain how we accommodated to this new goal as it developed over what were to be eight years of research.

One final remark. The reader will note that throughout this work we spell *Catalunya* as it is spelled in the Catalan language. In Castilian it is written as Cataluña. The *tilda* does not exist in the Catalan written language. The English spelling is *Catalonia* and in French it is *Catalogne*. We have chosen to use the Catalan spelling because it is phonetically correct and out of respect for those many Catalans who have cooperated with us in our difficult undertaking. Unless otherwise noted, all translations are our own.

Thanks are due to my editor at Palgrave Macmillan, Farideh Koohi-Kamali, and two Catalan friends, Maria Garcia i Valls and Pere Puig I Rossell, all of whom helped in improving this book. All errors that remain are, of course, my own.

CATALAN SPEAKING AREAS OF EUROPE

CATALAN POLITICAL MAP

Introduction

Prelude

[There is] an archetype of the Catalan that one can find anywhere, at any moment. In any kind of group—friends, fellow workers, and musicians—there are always one or two. These Catalans are easy to identify because when one talks to them in Catalan they respond in English (or, to Catalan speakers, in Castilian). In most cases, it just so happens, that my Catalan is much better than their English, so I continue to speak Catalan. Then, if they respond in Catalan, they do so in low voices as if they were ashamed to speak it. . . . If there is a pause in the conversation—they will switch back to English.
—Tree, *CAT*, 92

No vull dir que en Castellà	It's not repugnant to me
Treballi de mala gana	To write in Castilian.
Sino que em gusta molt mes	But I take much more pleasure
Lo llenguatge de la patria.	In the language of my country.

—Itgnasi Ferreres, eighteenth-century poet from Mallorca

I don't speak Arabic. My voice speaks the letters of the alphabet: a, ba, ta, tha. Then they are effaced. It's a hungry voice. It is a voice foreign to the language it speaks. I speak without understanding. This language escapes like sand falling through my fingers. It wounds me. It leaves its marks—certain words—and then disappears. It does not take me into itself. It rejects me. It separates me from the others. It separates me from my roots. It is an absence. I am an invalid. My land is stolen from me. I am here—different and French. But I am Algerian: by my face, my eyes, my skin. By my body traversed by the bodies of my grandparents, I bear the odor of their house. I bear the taste of their pastries and their bread. I bear the robes, the chants, the songs, the sounds of the clanging bracelets. I bear the hand of Rabia on my feverish face. I carry the voice of Bachir who calls his children. This voice hovers above everything. It sounds forth persistently destroying an emptiness. It is eternal and powerful. It binds me once again to the others. It makes me a part of the Algerian land.

—Nina Bouraoui, *Garçon Manqué*, 11–12

> Caliban: You taught me language, and my profit on't
> Is, I know how to curse. The red plague rid you
> For learning me your language.
> —Shakespeare, *The Tempest*, 54

Jacques Derrida speaks of his experience as an Algerian of Jewish origin who has lost his mother tongue, Ladino, the language of his ancestors, Sephardic Jews from Spain. Although monolingual and fluent in French he feels estranged from it. Derrida describes his malaise as follows:

> But above all, and here is the most fatal question: How is it that this language, the only language that this monolingual speaks, and is destined to speak, forever and ever, is not his? How can one believe that it remains always mute for the one who inhabits it, and whom it inhabits most intimately, that it remains *distant, heterogeneous, uninhabitable, deserted*? Deserted like a desert in which one must grow, make things grow, build and project up to the idea of a route, and the trace of a return, *yet another* language? (*Monolingualism*, 55–56)

> Today I shudder to think how a country, so foreign to my own instincts, could have achieved the miracle of being called mother.

> The English themselves were not aware of the role they had played in the formation of these black strangers. The ruling class was serenely confident that any role of theirs must have been an act of supreme generosity. Like Prospero, they had given us language and a way of naming reality. (Lamming, *In the Castle of My Skin*, xxxvii)

> The soul of a people is its language. (Goethe)

> If language is the criterion of ethnic identity, the disappearance of a language must also bring about the disappearance of its ethnic group.... It is only language that maintains the spirit of a national conscience. ([The artist] Ben, *Ben à Céret*)

> Gaelic in Ireland is called Irish, so that Irish people will remember what country they're living in. Some people say that the Irish language reminds them of the big famine when they had nothing to eat except the old poems in Irish. My father says people transformed everything they owned into English, their stories and their songs, even all their memories and their family photographs. They deny that Irish has anything to do with them anymore, but some of their ways of saying things come down from the old bards, even if they don't know that.

> "One day the Irish people will wake up and wonder if they're still Irish," he says.

And that's why it's important not to bring bad words like fruit gum into the house. That's why it's important to work hard and invent lots of new things in Ireland and fight for small languages that are dying out. Because your language is your home and your language is your country. What if all the small languages disappear and the whole world is speaking only one language. We'll all be like the Munster poets, he says, lost and blind, with nothing to welcome them only doors banging in the wind. We're living on the eve of extinction, my father says, One day there will be only one language and everybody will be lost. (Hamilton, *The Speckled People*, 161–162)

I soon understood that languages were closed worlds, that their translation could never convey the exact emotion of one word into another language. To say that the man in the port was dead, was simply not the same as saying he was *mort*, even if both words have the same meaning. The emotional connections between sound and meaning cannot be disentangled, for in doing so they are lost. In my experience "dead" was like a dull pain, like the quiet end of a smile.... *Mort* was the sudden tolling of bells, deep mourning, the whole scuttling up the hill to the church, a gloom beyond words, and the young men carrying the coffin on their shoulders, their hair plastered down with *brillantina*, the spotless Sunday clothes, the pride of their mothers or wives.

I spoke Majorcan, which is a variant of Catalan, with everyone in the village, and Spanish with people who were either from the mainland, or lived in Palma—where under the dictates of the Franco regime the Majorcan "dialect" had become almost relegated to the kitchen. (Graves, *A Woman Unknown*, 25)

Thirty years ago my wife and I bought a house in Southern France near the city of Montpellier. Over time we became friendly with several inhabitants of the village. One day we were invited to supper by a neighbor whose father, born in Catalunya and fluent in that language, had set out on foot from his village to seek work across the border in France. After finding a job he decided to remain in France and eventually started a family there. During a meal with this man's son our conversation turned to the subject of travel in other regions of France. Our host, whose last name is typically Catalan, but who speaks no Catalan, stated in astonishment and displeasure that when he had visited Strasbourg on vacation the previous summer he was surprised to hear people in the streets speaking to one another in what he took to be German. "After all," he said, "they live in France and should speak French." It is not uncommon in France to hear comments of this type even from those, like this individual, from an immigrant background.

Successive governments, beginning with Louis the XIV and continued in earnest by Napoleon, the "great unifier of France," created laws to suppress local (or regional) languages. Alsace is atypical, however. It has a long history of passing back and forth between France and Germany. After the Second World War, when it definitively became a part of France, it was granted a special exception from certain parts of the national legal system and its language was also officially recognized. Still as far as I know all, or most, Alsatians, are bilingual in French in addition to their own form of German. The vast majority of them are also famously loyal French citizens.

One day, quite recently, I heard a couple of tourists speaking French on Sixth Avenue in New York's Greenwich Village. They were obviously lost and I offered to help them find their destination, Washington Square. They accepted gratefully, and we set off toward the park. First they complemented me on my French, and then our conversation quickly turned to tourism and visits they and I had made to other countries. I mentioned that one of my favorite cities was Barcelona. The woman of the couple agreed enthusiastically, but added she thought it a shame that so many of the people there spoke Catalan, concluding with "They should speak Spanish. After all Barcelona is in Spain!"

These two anecdotes reflect a widespread feeling in France, the most highly centralized country in Western Europe, that regional languages need to be eliminated for the sake of national unity. If the American flag is the major symbol of the United States, French is the major unifying symbol of the French Nation. It was not surprising, therefore, that on December 22, 2003 during the 8:00 p.m. French news broadcast from the state network, France-Deux, I saw a highly laudatory documentary on the persistence of French language usage among the Cajuns of Louisiana. France had taken an active role in this preservation by sending language teachers from the home country to the Cajuns. This strikes me as hypocritical and contradictory given the French attitude to regional languages in their own country, an attitude that persists although, in the recent past, restrictions have been somewhat eased. There is no question that, even today, the French authorities, are strongly hostile to true bilingual education in the public school system. On the other hand such programs *are* tolerated in private schools.

Policemen on hearing anyone speaking Catalan on the street during the Franco regime were commanded to say: "Stop barking like a dog and speak Christian!" This response was not unique to Franco Spain, however. Well into the twentieth century it was commonly heard by schoolchildren in French Catalunya when their teachers heard them speaking in Catalan.

French is the language that guarantees that men cannot lie to one another. (Maurice Druon, writer [1977])

Above everything else, French is by its very nature, the language of human rights. Which language more expresses the essence of the human message "All men are equal."? No other language, no other culture so definitively expresses the dignity of each individual in the world, no matter who, no matter where. (Jacques Toubon [ex-] Minister of culture *in* Le Monde, January 7, 1994) (Quoted from Lluís Lluís, *Conversa Amb El Meu Gos*, 108–109)

But a language is more than an artifact. You can't slap a price tag on a language, no matter how small and obscure, any more than you can pin down the the financial value of an ivory-billed woodpecker or a bill of rights. (Abley, *Spoken Here*, 4)

On every inhabited continent, languages keep falling silent. New replacements are rare. Linguists believe that about six thousand languages still flow into human ears: the exact total is a matter of debate. By some estimates a maximum of three thousand are likely to be heard at the century's end, and fewer than six hundred of those appear secure. (Ibid., 4)

Advocates of Welsh like to look south—past the language's struggling cousin in Brittany, all the way to northern Spain. There, in the Basque country and Catalonia, minority languages are thriving, fostered by serious money from regional governments. ("We're like the Catalans," Sue Waters has said, "but ten or fifteen years behind.") (Ibid., 254)

If you live in Catalunya, work in Catalunya, speak Catalan, and want to be Catalan—you are Catalan! (A statement by Jordi Pujol, former president of the autonomous government of Catalunya.)

Chapter One

The How and Why of This Book

Anthropology is a peculiar subject. Unlike the other "social sciences" it lies at the crossroad of three intellectual pursuits: science, social science, and the humanities. Anthropologists engaged in fieldwork, although obliged to treat data as objectively as possible, must maintain a complicated set of social and personal relations with native informants. Thus, by the very nature of the anthropologist's task, a certain degree of subjectivity is unavoidable in the prolonged interaction process crucial to the pursuit of research. In the final analysis, most successful anthropological studies, while utilizing information as objectively as possible, are (in my opinion, in the best of cases) far from impartial. I choose not to work in a culture for which I feel a lack of empathy. Thus, I would argue that objectivity in anthropological research implies only that data appropriate to the project be collected and analyzed with care and in depth. What it does not demand is for the study's final results to be couched in value neutral terms. There is a tendency, however, among a number of my contemporaries to confuse the necessary relativism of value neutral analysis of a cultural practice, in order to understand it in the context of a particular society's history and social system, with a more extreme relativism—a relativism that refuses to take a positive or negative moral stand on a cultural practice. Honest understanding is not the same thing as moral neutrality.

In my long career as an academic, I have seen a dizzying flurry of theories come and go over a relatively brief time span: diffusionism, Malinowskian functionalism, Radcliffe Brownian functionalism, cultural ecology, human ecology, cultural materialism, structuralism, post-structuralism, structural Marxism, semiotics, postmodernism—and I have probably left out a few. This sequence reminds me of those professors in the recent Canadian film, *Invasion of the Barbarians*, who, toward the end of the story, ruminate over the plethora of theories that had captivated them briefly, each in its turn, during their academic careers. True, these fictitious professors were in the humanities but they might just as well have been in the social sciences. I have also noted during years of teaching the often uncritical enthusiasm newly minted graduate students eagerly displayed for the latest theory while rejecting the classic ethnographies (field studies of cultures around the world) as boring. What a difference in the hard sciences. If one compares the history of physics and biology with that of the social sciences a striking fact emerges. Newtonian theory was the keystone of physics at least until Einstein came upon the scene, and Einstein's own contribution continues to guide theoretical physics today. In biology, Darwin's theory of evolution has stood the test of time and has opened vast areas of research. Even the discovery of DNA and the recent advances in molecular biology can be linked to Darwin. What makes these theories so long lasting? A good theory should explain, and the explanation should be open to verification through further observation and testing. A good theory should also be designed to make predictions that move its domain forward. Newton, Einstein, and Darwin have survived the test of time because their theories meet these requirements. In contrast, the theories that have emerged in the social sciences are like foam on waves rushing toward the shore only to disappear in rapid succession. Although many such theories *have* limited value as research strategies, they are patently unable to make clear and successful predictions about future events. In fact anthropology's capacity to predict is about equal to that of astrology.

Part of the problem, of course, lies in the nature of human behavior loaded as it is with complexity and unforeseen circumstances. In general, this makes it impossible to isolate all the relevant variables and perform verifiable experiments under laboratory conditions. When I taught my undergraduate introductory course at Columbia, I was fond of beginning each semester by saying that anthropology is good at predicting—the past. So what's left? The answer is careful, persistent, long-term research and testable interpretation involving verification of specific large- and small-scale recurring events and attitudes. As my graduate advisors at Yale told me long ago, a good thesis needs only to be a careful analytic description of a specific culture.

Today one of the most popular "theories" in the humanities and the social sciences is postmodernism, which, in my opinion, is actually antitheory, but one that, ironically, has its own theoretical bias. I see much of the postmodern stance as little more than a modern replay of the subversive Dada movement in the arts that briefly came upon the scene just after the First World War, and that challenged accepted ideas and prejudices in the arts and society at large. Let me make it clear that I find most of what postmoderns have done to be as destructive as any other failed theory. What then are my objections? For example, as I have already noted briefly above, one of the dogmas of postmodernism is an exaggerated view of the place of cultural relativism in research and analysis. Relativism in anthropology is an approach first advocated by Franz Boas, the father of American anthropology. For Boas it meant that when one engages in anthropological research one should remain value neutral and gather the data no matter how bizarre or unpleasant it might be. Put another way, researchers in the field situation should never let their own prejudices and value judgments get in the way of data collection, nor should such values guide their conclusions.

Many postmoderns have a sometimes hidden political agenda (there are exceptions) based on a strong bias against the status quo in our own culture. This leads to the claim that any and all standards of interpretation or judgment are based purely on social conventions and are, therefore, impossible to verify in any objective sense. Thus, to give one example, it makes no sense to say that Shakespeare was the greatest writer that the English language has yet produced. The postmodern response is that the evaluation of Shakespeare is a reflection of a currently accepted standard based solely on subjective criteria linked directly to the existing power structure in our society. The next step in this line of reasoning is to extend it beyond aesthetic judgment to the claim that modern science is merely one paradigm among many possible paradigms and has no more objective validity than any other theory or method.

Another principle of this form of cultural relativism is that no one has the right to criticize the practices of a culture that is not one's own. In my last year of teaching the undergraduate introductory course in cultural anthropology at Columbia University, I assigned two articles dealing with what is euphemistically called "female circumcision." This operation occurs in parts of Africa and is believed by practitioners to be a dictum of Islam (which it is not). It involves different degrees of genital mutilation that can go so far, in some cases, as the total excision of the clitoris of young girls. When I asked students in the class to express their opinion of these articles and the practice itself, I was astonished to find that most of them, including women, were of the opinion that we, as strangers to the cultures that perform this operation, have no right to criticize it. If it is someone else's culture, they said, we must remain value neutral!

The logical outcome of the above arguments is that anthropology is, at best, a form of writing akin to fiction and that no one can possibly understand, describe, and/or accurately analyze another culture. Therefore, ethnographies are merely texts about other texts, the latter being the data of anthropological research written down in the anthropologist's notebooks. I know of a famous anthropologist who has gone so far as to claim that what he now does is "ficto-criticism," a new genre of anthropological writing rather than traditional anthropology. His area of specialty is the country of Colombia. In my opinion ficto-criticism, certainly in this case, should be left to Gabriel Garcia Marquez who does this type of writing with exceptional brilliance and is, of course, a native informant as well as a great literary talent.

Can One do Anthropology Well?

A confession: I admit readily that my first fieldwork was under par, if still good enough to get my PhD and move on. But I can see that the frailty of my thesis, when I look back on it, was the result of a much too short field stint combined with a poor grasp of the native language rather than some fault inherent in the discipline. Now, I am not so naive to think that one can collect valid and insightful data without some notion about a methodology that is appropriate to the subject of the study: what one should focus on and what to ignore in that massive amount of data that is the life stream of the cultural group under investigation. This is not always a decision easily made. I have been fortunate in recent years (because of rank and tenure) to have unusually long periods of time to complete my last two field projects and, therefore, to decide what should be the best methodology to employ and to slowly correct the focus of research as the work progressed. One needs to check and recheck facts and figure out what really counts in the stream of everyday and, also, exceptional behavior. Anthropological analysis demands a great deal of patience.

Finally, many, if not most, postmoderns write badly on purpose! I believe it is their intention to terrorize the reader either into submission, or at least to admit that the author is much more intelligent than the reader. These postmoderns hide their lack of interpretive power under impenetrable academic prose. In contrast I am convinced that there is nothing in the social sciences that cannot be expressed in clear English. I try very hard to write clearly. Whether or not I succeed is, of course, up to the reader to decide.

Why Catalunya? A Bit of Personal History

Both my parents were typical immigrants in the great wave that occurred at the end of the nineteenth century up to the first quarter of the twentieth century. My mother arrived in the United States from the

Ukraine in 1906, with her parents and two sisters, in order to escape anti-Jewish pogroms that were common at the time. She was only four months old when she arrived at Ellis Island. Her extended family, however, had already established roots in New Jersey where my mother's grandfather, the family patriarch, along with his wife, had arrived some years before with some of their eleven children. Anxious to become Americans my mother's grandfather, and later her own parents, eager to learn English, soon banned Russian as well as Yiddish from family discourse, except to use one or the other language to keep secrets from the children of the household.

My father's history is somewhat different. He lived with his family in Sevastopol in the Crimea where his father was a successful factory owner. According to Russian law, Jews were not supposed to live there because it was an important strategic naval base. My father's family was excused from this restriction because his grandfather had been decorated as a hero of the Crimean War. My father left home because of the Civil War that followed the Russian Revolution. He first emigrated to Turkey where he lived for three years and then decided to move on alone to the United States where he had only one distant cousin. Russian, of course, was my father's native tongue, but his family spoke French around the dinner table, a frequent practice in middle-class Russian homes at that time. Shortly after arriving in New York he moved to New England, on the Quebec border, to find employment in a pencil factory where many of the workers were French speaking. It was there that he began to learn English, perfected when he returned to New York City a few months after his arrival in the United States. Although he later taught my mother Russian, the language of our household was English with only occasional lapses into Russian, again used as a means of keeping secrets from their son.

I was a poor language student in high school and, in college, avoided taking language courses. In 1967, my wife Sonia (who had majored in French language and literature) and I bought a house in Southern France. It was there, by necessity, that I began in earnest to learn French. We spent two sabbatical years in Paris in 1968 and 1973. During the latter year, my French language skills improved such that I was able to converse comfortably, attend lectures at the Collège de France, and join Professor Claude Levi-Strauss' seminar at the École Des Hautes Études en Sciences Sociales.

My interest in Catalan identity, the subject of this book, is tied in part to the sense of loss I feel in regard to the Russian language. In contrast to my parents' refusal to teach me Russian, the Catalan people struggled through three periods of Spanish history to keep their language alive in the face of a ban imposed by centralist governments in Madrid. The most recent of these events spanned Franco's dictatorship from 1939 until his death in 1975. In 1977, the autonomy of Catalunya was reestablished and the Catalan language was officially recognized as coequal with Castilian. Today it is the first language in both elementary and high school. On the university level, professors have the right to teach their courses in either Castilian or Catalan, a source of conflict within Catalunya that is discussed a little later.

Although I was drawn to Catalan culture, there were other reasons for choosing this study. My interest was also sparked by research I had previously completed in France. This concerned a small group of farmers on the Larzac plateau located about one hundred kilometers from the Mediterranean coast and about thirty minutes by car north of our house. These farmers successfully struggled for ten years against the expropriation by the central government of their farms in order to extend a military base from 3,500 to 17,500 hectares. The struggle—lasting from 1971 to 1981—began at a time when the highly centralized French government had just experienced an ideologically similar, if unsuccessful, upheaval in Paris led by university students allied with the working class. This failed protest, however, later led many of the participants to join the farmers' movement in the south that was based on a different strategy. The major difference between these two confrontations with state power was the central ideology of the farmers' movement, which offered a new means of struggle, based on nonviolence, and modeled on the principals of Gandhi and Martin Luther King. Where violent confrontation had failed in Paris, nonviolence endured. This successful struggle, and its still active aftermath, was the inspiration that formed my interest in the culture of conflict between center and periphery in the context of the nation-state centered around nonviolence as a means of action. Catalan nationalism (many Catalans prefer to call it "Catalinism") is, in fact, nonviolent, although the history of Catalunya is replete with violent confrontations between Catalans and the attempt of foreign powers to conquer their territory. Catalan's recourse to violence has been, throughout history, mainly defensive.

Another attractive aspect of Catalan identity, I soon discovered, is that it is based on a person's active knowledge of the language and not on heredity. This distinguishes it from the common form of contemporary nationalism that is racist in origin. In fact, as George L. Mosse points out, the link between nationalism and race so common today dates only from the end of the nineteenth century and that "racism was never an indispensable element of nationalism" (163). Does this mean that there are no racists in Catalunya? Of course not. If no person is perfect, neither are cultures immune to imperfections.

How This Study Came Into Focus

At the end of our French study, on sabbatical leave in 1991–1992, I turned on the regional television station (*France Trois*) to watch the local news. A feature story caught my eye. It concerned an attempt by two mayors of adjacent communities—one, Cerbère, in French Catalunya, the other in Portbou just over the border in Spanish Catalunya—to develop a transborder cooperative project as a reaction to the coming of an open border between the two countries as members of the European Community; hence the disappearance of the customs bureau in each village. On the strictly local level, this change was a major threat to the economy since the customs clearance business (the employees of such companies were called "transitors") was an important source of employment in governmental and private businesses.

Since both villages were, in principle, if not in fact, in the contemporary situation, Catalan by culture, the mayors hoped to capitalize on cultural ties between them to encourage tourism and, if possible, new business opportunities. One major project was to build a new school on the border between France and Spain. To build the school, the mayors required the approval of two governmental bodies. Portbou, because it is part of the autonomous region of Catalunya, only needed the permission and help of the Catalan government in Barcelona, which generally supports such initiatives and which readily agreed. The situation in Cerbère, though, was quite different. Approval in this case depended on the highly centralized government in Paris and the answer from there was a definitive no! This decision, however, came after the news broadcast to which I had responded with great enthusiasm. Here was a situation, I believed at the time, in which a single culture with a common heritage and economic interests could provide the means for strong cooperation across national boundaries.

Although our project, as we later discovered, was based on a false premise, it turned out to have other virtues for a comparative study of cultural and national identity. Both Cerbère and Portbou were founded in the same year, toward the end of the nineteenth century and for the same reason, which was essentially to overcome an existing discrepancy between track-width in the national railroads of the two countries. The space between tracks in Spain is wider than in France. As a result of this incompatibility commercial trade between the two countries required that products be transboarded at each railhead. Originally this was done by hand, an exhausting job, generally performed by women laborers. Recently a modern technological system has replaced the older method. This involves the mechanical transfer of entire box-cars by huge cranes from Spanish to French wheel systems, and vice versa.

Another traditional, if illegal, link between the two towns was smuggling—a practice that considerably enriched certain inhabitants on both sides of the border. The rugged coast in the area provided easy cover for the transfer of illegal goods away from the prying eyes of police and customs agents. Such trade was, of course, also facilitated by the common language and culture shared by citizens living in each village. Smuggling persisted well into the Franco era, and even later. Attempts to control it required the heavy presence of Spanish and French gendarmes whose charge was to patrol the border adding significantly to the population and hence to the economies of both villages. Small-scale smuggling between Cerbère and Portbou is colorfully described in the novel, *La Casa Gran*, by the Catalan novelist and Portbou native, Maria Mercè Roca, whose father was a small-scale smuggler during the Franco years.

Getting Started

My original research plan was to focus exclusively on Cerbère and Portbou as a natural laboratory for comparing the status of Catalan identity and language on both sides of the international border—thus one culture, two nation-states. By the academic year 1996–1997, I had accumulated another yearlong research sabbatical and was ready to begin. My wife and I chose to start work in Cerbère because we had not yet learned Catalan. Our French was more than adequate for the first phase of the research. Catalan language study was to begin later, during the winter of 1997, when we had arranged intensive language classes with two private teachers. Our first stay in Portbou was delayed intentionally until the spring of 1997 so that we could work exclusively in Catalan.

Cerbère

Cerbère is located at the end of a road that begins near the resort towns of Argelès with its wide flat beach area and Elne with its lovely

Romanesque church and cloister. After Elne, the road begins to wind upward on the rise of the rocky, schistous promontory known in France as the *Côte Vermeille*, the "Ruby Coast." It leads to access roads for the popular seaside tourist towns of Collioure (where the painters known as "Fauves" established themselves for a short time at the beginning of the twentieth century). After Collioure, a now improved road arrives at the busy shipping village of Port Vendres; then the road becomes narrow and twisty on its way over a few more hills to Banyuls, a center of local high quality wine production and an important seafront tourist site. After Banyuls the road becomes even narrower: a precipitous twisting route with over one hundred sharp curves as it winds toward its final destination in France, Cerbère, only six kilometers away!

Cerbère is built around a narrow rocky beachfront protected by a break-water. In summer and in fall a floating port for pleasure vessels (taken up for storage at the end of the season) provides docking space for small motor- and sailboats. Cerbère has only three hotels (two of them quite small and the other only medium-sized), a range of café-restaurants, and various shops, many of which cater to tourists. Cerbère's neighbor, Banyuls, is the headquarters for a protected underwater nature reserve that stretches all the way to the border with Catalunya, but this facility is strictly for research and divers are not permitted to enter it. Although from its founding days Cerbère's major business activity centered on the railroad, customs clearance, and the transboarding of produce between trains, it also has an agricultural side. Again, because it is close to Banyuls famous for its strong apéritif wine (similar to sherry) Cerbère is included in the restricted area of the prestigious Banyuls appellation. The vineyards here are relatively small and rise on terraced rocky slopes that cling to the sunny side of the steep hills, which is characteristic of this coast. Finally, Cerbère is the home of an important medical center, Peyrefite, for the treatment and rehabilitation of severely handicapped stroke and accident victims. Although this facility provides employment for a small number of unskilled workers inhabiting the village, the majority of the staff is specialized and commutes from as far away as Perpignan. Today, although the beachfront is well cared for and lively, the rest of the village is somewhat banal and run-down in appearance. This is the result of economic changes that began after the Second World War with the development of the mechanized transfer of freight between the differing Spanish and French wheelbases.

Just beyond the beach a short river, dry in summer, but often a raging torrent in winter, is lined with houses raised on a cement platform constructed to contain the water below. There are also a few shops on this short street, many of which are closed due to a severe fall-off in business by the time we began our research. The rest of the town rises steeply from the beach to houses perched on rugged hills. A large apartment building constructed to house railroad workers is built on one of these rises. The most striking feature of the village, however, is an architecturally unique building, the Belvedere Hotel. This edifice, a product of the Art Deco period, resembles a great ship cast up from the sea that has been dropped onto the rocky promontory above the water. To add to the image of a wayward ship, the hotel is built over the railroad tracks that pass directly under its massive hulk. The Belvedere had its heyday during the 1920s and 1930s, and briefly in the post–Second World War period, when its guests included movie stars and important political figures. The hotel used to house a cinema and a gambling hall but today only few of its rooms are occupied as long-term studio rentals. The overall impression of the building is one of abandon—run-down and forlorn. Still, it is a monument to be seen: a marvel of its time.

Cerbère, squeezed as it is into a narrow space, with its small rocky beach, has little of the charm of the other towns on the Ruby Coast, and many who visit the village either come for short stays or stop there briefly on the way into Spain. Just as Portbou is a point of entry to Spain, Cerbère is a point of exit from France. For years during the Franco regime, and even afterwards, the French franc was stronger than the Spanish peseta making it advantageous for tourists to go southward for their vacations. Even today, with the euro as a common currency, Spain is somewhat cheaper than France due, at least in part, to the fact that the added value tax applied to hotels and restaurants is considerably lower there.

We Begin Work

Arriving in the early autumn afternoon, it did not take long to find a small apartment conveniently located in what was essentially a made-over garage with a balcony overlooking the sea. Once settled in we began to introduce ourselves around the village. In order to get to know people, we did not hesitate to start conversations with strangers in cafés, various grocery stores, and shops selling tourist items. We also went to the town hall to meet the mayor and were cordially received by him as well as by his staff. In general, people were friendly, and we were soon able to create an expanding web of acquaintants.

A few of our new informants introduced us to some key people in the village beginning with the secretary of the local tourist office, who like some other young people in the village, was born in Portbou but had married a Cerbère native and settled there with her husband. One of the café owners, a Catalan speaker (but by no means a Catalan nationalist) introduced us to one of the very few Catalan militants in the village. He also introduced us to a local storekeeper who like himself had recently been elected to the town council. From then on things were easy. We were able, during our short stay, to interview the newly elected mayor, other council members, a range of shopkeepers, two active members of the railroad union, and the oldest couple in Cerbère, important to the village as the unofficial keepers of village history. We spent a day in the Cerbère elementary school where we sat in on a class and interviewed three teachers, two of whom were Catalan speakers. As promised, we returned to the school the next spring to talk about and show slides of New York to the children.

And so it went throughout our stay. It did not take us long to develop a widening group of informants, one of whom even offered help finding people to interview in Portbou where he had many friends and relatives. An additional lucky break during our stay was the discovery of an unpublished French PhD thesis in political economics concerning the history and economic development of both Portbou and Cerbère from their founding at the end of the nineteenth century up to the 1980s. I later obtained a copy of the thesis. Although our stay in Cerbère was short it was productive and laid the groundwork for further visits through the next several years.

Language Study in Barcelona and First Impressions

At the end of our two months in Cerbère we left for Barcelona to begin our language training. We rented a room (including board) for three months in the apartment of a woman fluent in both Catalan and Castilian. We found her, as well as our future language teachers, through a Catalan who had participated in the Larzac struggle. Our landlady was proud of her Catalan heritage and friendly with several important Catalan cultural figures to whom she was kind enough to introduce us. We were lucky to find her since most Catalans are loath to invite even paying guests into their homes.

During our stay in Barcelona we spent four hours a day, Monday through Thursday, taking language classes and, in the afternoons, studying for the next lesson. On weekends we took side trips to the Catalan countryside to see how the language was fairing outside of Barcelona. In that city Castilian is widely spoken, particularly by members of the working class. (Approximately 50 percent of the population is composed of immigrants from other Spanish provinces as well as Latin America, and are not native speakers of Catalan.) It soon became apparent that most people in low-paying service jobs were, more often than not, monolingual in Castilian. We also discovered that many of these monolingual speakers of Castilian were loath to acquire Catalan for a variety of reasons, ranging from the supposed difficulty of learning a new language (not very severe for native speakers of Castilian) to those frankly hostile to Catalan culture. Additionally, we found out that some Catalan-speaking parents send their children to private, usually Catholic, schools where, at least sometimes, Castilian is favored over Catalan. During rides on the Barcelona metropolitan subway system we rarely heard teenage passengers speaking Catalan among themselves.

The press in Barcelona publishes largely in Castilian although a strongly Catalinist paper, *Avui* (Today), founded shortly after the death of Franco, appears daily—exclusively in Catalan. This paper gives voice to a wide range of opinions (expressed in interviews and guest editorials). From time to time, these include the opinions of members of the *Partit Popular de Catalunya*, a local version of the national Popular Party, right wing and generally hostile to the linguistic policies of the Catalan government. (One of the ironies of the current Spanish government is that while it attacks the Basques and Catalans as nationalists, it denies that its own point of view is nationalist.) Recently one other paper, *El Periòdico*, at first published exclusively in Castilian, began to publish in a Catalan edition. The regional Catalan paper, *El Punt*, now also publishes a Barcelona edition. Another voice for Catalan language and culture, including its expression outside of Catalunya proper (the Balearic Islands, València, the independent country of Andorra, and Roussillon, the historically Catalan part of Southern France) is the magazine, *El Temps*, published weekly in the city of València. We have followed events in Catalunya as reported in *Avui* since 1996, and since 2002 in *El Temps*.

Although a wide selection of Castilian (nationwide) TV stations are broadcast throughout Catalunya, the listening public also has access to three Catalan stations: the popular TV 3 and Canal 33, the latter offering a greater range of cultural programs than its sister station. In the fall of 2003, a new TV station, reporting the news twenty-four hours a day, began broadcasting.

As for Catalunya radio (operated under the auspices of the Catalan autonomous government), it broadcasts exclusively in Catalan on four stations: Catalunya Information, Catalunya Music, Catalunya Culture, and the more general Catalunya Radio. There is also a wide choice of private stations in both Castilian and Catalan. During broadcasts on Catalunya Radio dealing with Spain-wide questions (the actions of Congress in Madrid for example) and also interviews with monolingual Castilian guests, the conversation is often bilingual with the Catalan host speaking Catalan and the respondent answering in Castilian. In cases like these no translation of the Castilian into Catalan is offered. It is assumed correctly that the majority of the audience will at least comprehend the responses in Castilian. The same is true for Catalan programs broadcast on TV. All, or almost all, Catalan speakers are bilingual in their own language and Castilian. This is one of the reasons many of them resent those Castilian speakers who live and work in Catalunya and refuse to learn its language. Early in our stay in Barcelona we had an experience with this attitude when a leak developed in the kitchen of our landlady's apartment. The plumber who came to fix it was a Castilian speaker who did not speak any Catalan. When Sonia asked him whether he spoke Catalan, he replied, "Why should I! I have been here for thirty years and have no need to speak a language other than Castilian. Besides we are in Spain!"

In addition to the mass media, a large number of political and intellectual publications (the domain of the "small" press) exist in the Catalan language. Also of note is the fact that well before the death of Franco, the Abbey of Montserrat, a center of resistance to fascism, was the first institution to publish and distribute books in the Catalan language. Initially, these were exclusively religious texts, which the Falange government would have been embarrassed to forbid. Later, the United States, wishing to establish military bases in Spain in the 1950s and keep Franco's Spain free of communist domination, put pressure on Franco to ease both censorship rules and allow a certain number of books and magazines to be published in Catalan by commercial publishing houses.

It was our strong impression in Barcelona, and later in other regions of Catalunya, that real bilingualism occurs in native speakers of Catalan considerably more than in the immigrant population. Additionally, we soon discovered that in social situations including both bilingual Catalans and Castilian speakers, whether or not the latter also spoke Catalan, the Catalans present would switch to Castilian.

Portbou

Although our Catalan was far from fluent after three months of language training, we felt comfortable enough in it to move on to Portbou where all our conversations and interviews were to be held in that language. We chose this strategy in order to demonstrate our interest in Catalan culture, the raison d'être of our study. Bilingual Catalan speakers of Castilian, even those of Castilian origin, never questioned or indicated any negative feelings regarding the use of Catalan in our interactions with the local population. In addition to carrying out interviews on our own, we hired a bilingual Catalan to conduct a questionnaire with villagers a few of whom turned out to be monolingual in Castilian. In 2003 we asked another bilingual person to conduct a supplementary number of questionnaires.

Portbou is located about three kilometers from Cerbère by a narrowing winding road that runs across the border over a mountain pass to drop down to the village perched below on a small rise just beyond its rocky beach. It is surrounded by schistic cliffs typical of this coastal region of Catalunya known as the "Costa Brava." To reach the next inhabited village (Colera) one must once again spiral up a mountain and then descend to the other small village. Beyond Colera one finds Llancà, much larger than Colera or Portbou, with its own ration of tourists during the summer season. Llancà boasts several large beaches, numerous hotels, and a large port for pleasure boats. Next on this rocky coast is the lovely and quieter Port de la Selva, an active commercial fishing station. This village of whitewashed houses rises abruptly from the seaside.

Port de la Selva is located on one side of the peninsula "Cap de Creus." Big-time tourism begins on the other side in Cadaqués, Salvador Dalí's home town. Here one finds the first sand beach on the rugged coast and a number of medium-sized hotels. Beyond Cadaqués lies the summer destination of Roses, with its hotels, restaurants and a crowded beach. In the summer, Roses is visited by waves of international tourists. In my opinion, it is a place to be avoided in any season, although recently a restaurant opened there that is noted in the prestigious French guide, (Le Guide Michelin) as serving the best and most innovative food in all of Europe.

Portbou itself looks in two directions: toward its seafront shops and restaurants and away from the sea, upward toward its railroad station. After a few days in the village one comes to realize that its only noteworthy source of employment beyond its shops and cafés is the Spanish rail system (RENFE). (The wine industry never took root here as it did in Cerbère although the climate and physical situation of the two villages are very similar.) There is, however, one small, but busy, boat-building and carpentry enterprise, the owner of which also runs a marine supply shop in town. This is Mr. Centellas, one of the village notables. Tourism in Portbou is limited to a very short-staying set of visitors, primarily French, and some vacationers from Girona and Barcelona, the latter only two and a half hours away by car and three by train. Of these many have, or had, family ties in Portbou. The majority of the French tourists pass only a few hours in the village, enough time to buy *pastis*, the licorice flavored apéritif popular in France (a cheap loss-leader in the Portbou shops that cater to tourists), perhaps spend a few hours on the rocky beach, and finish the day in one of the seaside restaurants eating a fast food version of paella. There are only two medium-sized hotels in Portbou (each with fourteen rooms, and two "hostals"—the latter offering simple accommodations for reasonable prices). In the winter months, only a single hotel on the seafront remains open.

Porbou's situation on the mountainside limits urban growth even more than in Cerbère. There is little space for construction, at least in areas that have a sea view and are convenient when compared with the narrow lower town. It is for this reason, and owing to a growing population of retirees, that real estate prices are unusually high when compared with Llancà and other towns and villages further south.

The rocky beaches in the village (there are three) are small, but the water, immaculate. Four café-restaurants line the beachfront. The town also has a short but picturesque "*rambla*" (a paved area for strolling) shaded in the summer months on both sides by closely planted and well pruned plain-trees that provide cover from the hot sun. Just beyond the rambla a tunnel under the rail line, often flooded in the winter months, offers access to land on both sides of a road that runs beyond the village proper. Here one finds a few houses and a number of vegetable gardens. About a mile beyond the tunnel, the road ends at the base of a dam that rises to an artificial lake that supplies water to the village.

Except for a few shops along the *rambla*, and those located a few blocks upward from the seafront, commerce beyond the immediate beach area is limited to a few short streets. The only other main shop and café-lined street is on a hill that plunges downward directly from the path to and from the railroad station. Much of the business on this street involves people arriving or departing by train. The village indoor market is located further down the same street, which also harbors two banks, a tobacco shop, and a post office. Numerous small cafés, some of which in the off season are frequented primarily by Catalan speakers and others by Castilian speakers, are dotted throughout the village. There is, however, no real social segregation in the village (no neighborhoods exclusively Catalan or Spanish).

Dwellings, most of them single or two family houses, border the streets that rise into the upper reaches of the town. Pedestrians can climb these hills via several daunting staircases that rise precipitously upward and that lead ultimately to the astonishingly large and impressive railroad station with its arching glass roof over the rail platforms. The station is the architectural glory of old Portbou and would not be out of place in the center of a major city. It is the last stop on the Spanish side of the border before trains going north plunge into the tunnel that reduces the trip to Cerbère to a matter of only a few minutes. This station, still active at this writing, stands as a striking witness to the past commercial glory of the town. To the right, as you face it, stands a large (late-nineteenth-century Gothic) church. This impressive building provides another sign of the financial wealth that Portbou once boasted as an important commercial point of transit between two countries. In both Cerbère and Portbou older people spoke to us of "the good old days" when businessmen "lit their cigars with paper money," the image borrowed, perhaps, from old American films.

There is one other unique and important monument in Portbou, this one of recent origin. It is a memorial to Walter Benjamin, the German-Jewish literary critic and philosopher, who committed suicide in the village when the Spanish National Police refused to honor his transit visa from Nazi-occupied France that would have allowed him to cross Spain into Portugal and eventually emigrate to the United States. This monument, an inspiring work by the Israeli architect, Dani Karavan, was financed by the German government. Benjamin's remains now lie somewhere in an unmarked pauper's grave. In addition to the Karavan monument, however, a memorial stone placed between rows of tiered niches, typical of Spanish graveyards, now marks his sad and short stay in the village. A group of people in Portbou interested in Benjamin's life and scholarship, in cooperation with a group of German intellectuals, worked for some years to create a center for Benjamin studies in the village, and in 2001, the Walter Benjamin Foundation was established. The opening ceremonies were held in Portbou's large church, and attracted a significant group of international scholars and Spanish political figures as well as local residents. The foundation as originally planned was still-born and an effort is now underway to revive it under different sponsorship, possibly to include the now ex-president of Catalunya, Jordi Pujol, who on finishing his term of office in the fall of 2003 expressed his personal interest in becoming the foundation's honorary president.

On arrival in Portbou we took a room in the beachfront hotel and set out to introduce ourselves and our project to the mayor, a former star of the Barcelona football (soccer) club, and a native Castilian speaker. This (now former) mayor is married to a Catalan woman from the city of Girona where speaking Catalan is the norm for a major part of the population. When we met him, the mayor was in the process of learning Catalan, but was still clearly more comfortable in Castilian. He welcomed us warmly as he struggled to speak Catalan with us. After describing what our study was about, we mentioned our need to find lodging in the village. Luckily, it turned out that his wife owns several apartments in the village and had one available conveniently located in the center of the village. Later, after we had settled in, the mayor kindly invited us to attend public meetings of the Village Council.

We accepted his offer for what turned out to be a totally frustrating experience, a foretaste of the confusion created by the bilingualism of Catalunya. At that time meetings of the Village Council turned out to be a blur of words, a fast barrage of language(s), half in Castilian, half in Catalan, with most of those present speaking with lightening speed and, often, all at the same time. Sessions such as these were incomprehensible to us. In the summer of 2003, during one of our annual returns to Portbou, we were once again invited to attend an important Village Council session. In spite of the importance of this particular meeting and the rancor it inspired among majority councilors toward the minority councilors, it stood in stark contrast to what we remembered of our first experience. Except for two relatively short outbursts in angry Castilian from the mayor, who speaks Catalan well but as a second language, the session was held entirely in Catalan with council members delivering their generally polite remarks, one at a time!

Our large furnished apartment overlooked the beach and the sea beyond. We quickly settled in and began to make the rounds, establishing contacts with villagers the same way we had in Cerbère. It was clear at once that the village was in the midst of the economic crisis similar to that of Cerbère and discussed earlier in the chapter.

As it turned out, even the railroad as a source of economic relief was endangered by a plan to build a high speed (TGV) line to run from Perpignan in France to Barcelona in Catalunya, thus bypassing Portbou. Though the project at the time of writing is still years from completion, when finished, it will deprive Portbou (as well as Cerbère) of a major part of their passenger traffic, including tourists. Since the announcement, a strong effort has been made by the French and

Spanish rail companies to compensate for the expected reduction in passenger traffic by increasing the amount of commercial goods transported on the existing system. To facilitate this plan the SNCF (the French national railway) has lent one of their traffic directors to the Spanish RENFE. Though an increase in commercial traffic would indeed help the rail lines to maintain the present level of employment, tourism would undoubtedly continue to decrease with the disappearance of passenger service. The alternative, tourism by automobile, had, since the 1950s, already replaced much of the long distance passenger traffic, but access to Portbou from other points in Catalunya, although not as difficult as travel by car on the French side of the border, is still painful due to the condition of the road that divides Colera from Portbou.

Thus, a possible solution to the present economic difficulty would involve projects to improve the automobile link between Banyuls and Cerbère, as well as between Cerbère and Portbou. Southward the Spanish government has already agreed to improve the road between Portbou and Colera with a tunnel that would shorten the time for automobile traffic between the two villages to a few minutes. Beyond Colera southward the roads are reasonably straight and safe. People from Portbou seeking employment outside the village, but desiring to keep their residence in the village, would have an easy commute to the city of Figueres where it is not difficult to find work.

On the other side of the border, however, while the French government has already invested in improving the road from Perpignan on the west to the *Côte Vermeille*, it has shown little interest in continuing the project beyond Port Vendres, the town just before Banyuls. Neither France nor Spain has even suggested that they might want to make the cross-border access between Cerbère and Portbou easier, although the mayors of both villages would welcome such a project.

Aid to Portbou was not linked exclusively to the coming tunnel project that is subsidized by the central government in Madrid. Even before our work began, the Catalan government (the *Generalitat*) in Barcelona and the European Community had made considerable investments in the village to aid the construction of a new port facility for pleasure craft in Portbou. Although local ecologists opposed the port on the grounds that it would spoil the natural charm of the site, the project was brought to completion in the summer of the year 2000. Though the port is likely to produce more tourism by boat and generate some funds for the village coffers, it is, as of this writing, unclear whether or not it will contribute to the reduction of unemployment in the village since it is not a labor-intensive enterprise. The *Generalitat* has also spent a tidy sum to upgrade and convert an architecturally magnificent private mansion into a community center that now houses an exhibition space, a library, and a center for retired people. Most people in Portbou, however, agree that this facility, stunning as it may be, is underutilized.

Additionally, when the project to build the trilingual school on the border fell through, the *Generalitat* financed the building of a new school in Portbou. Unfortunately, however, for economic and demographic reasons, the educational authorities in the region recently decided that students in the upper grades would have to commute to Llançà to finish the last phase of their elementary education. Many Portbou parents were upset by this decision. They were particularly bothered by the dangerous curves on the road between the village and Colera. In response to the announcement, they organized a well-attended protest at the railroad station where they blocked an express train on its way to France. As a result, the children concerned were allowed to split the week between school days spent in Portbou and those in Llançà. The new tunnel will certainly contribute to the calming of village sentiment concerning the safety of their children, but most of them would be happier if the students could finish all of their elementary education in the local school to which they are emotionally attached. Unlike French public schools where teachers and directors reign supreme under the watchful eye of the central government and local political authorities, parents in *Catalunya* play a key role in running their local schools, even selecting school directors.

Another controversial project recently proposed involves the construction of more than thirty electro-generating windmills (at this writing reduced to seventeen much larger and powerful ones) to be located on the heights surrounding the village. Although some opposition to these structures has been expressed by local ecologists who feel the windmills will denature an important natural site, when we left the village after a short visit in the summer of 2002, they seemed about to materialize. From a purely economic point of view the windmills (to be built by a private enterprise) will, no doubt, supply a viable source of income for the village in the years to come.

Demographic Similarities and Differences Between Cerbère and Portbou

During their years of economic prosperity both villages had populations of over 2,500. Since the border opening they have each lost about half of their inhabitants. More serious has been an increasing imbalance in the demographic pyramid. At present about half the population in both villages is made up of retired people. The major cultural-demographic difference between the two villages is the proportion of Catalan speakers versus Castilian speakers in Portbou and between Catalan speakers and French speakers in Cerbère.

About half the population of Portbou speaks Catalan. Of these, close to 100 percent also speak Castilian. Among the Castilian speakers some are monolingual, but many speak at least some Catalan as well, and others are fluent in both languages. Of those bilinguals who consider Castilian to be their mother tongue, the language of the household is most often Castilian. In mixed marriages frequently one language or the other is spoken with the appropriate parent.

All children of the village who have been educated in the village school rapidly become fluent in Catalan since it is the standard language of the school. This even applies to those who begin school as monolingual Castilian speakers. But, this does not mean that all of them use Catalan in social situations. In fact, there is a saying all over urban parts of Catalunya that while children speak Catalan in the classroom, they switch to Castilian in the school playground. There are many middle-aged and young people in the village who use both languages, Castilian with one set of friends and Catalan with another set. As is the case throughout much of Catalunya, those who are accustomed to speak Catalan with other Catalans will switch to Castilian if they know that in a group there is even one person who is monolingual in Castilian, or even if they perceive that an individual who knows Catalan prefers to speak Castilian. The exceptions to this rule are those who are strong Catalanists (see chapter four for more detailed information on the linguistic situation in Portbou). Since most of the restaurants in Portbou cater to tourists, menus are often in French, German, English, and Castilian. Few post Catalan menus although the law requires them to do so.

It would be impossible for a casual visitor to Cerbère to know that some of the natives of that village speak Catalan. The only language heard in public is French. The only outward signs of Catalan identity in Cerbère are that its restaurants offer some Catalan specialties and the distinctive Catalan flag is displayed on a few shops and houses. Also, the sign indicating the name of the town, following the general rule in the department, is bilingual: Cerbère/Cerbera de la Marinda.

The Two Villages in the Light of Tourism

For tourism Portbou has, in the past, maintained an advantage over Cerbère, one, however, that should not be exaggerated. For many years before the death of Franco, and years afterwards as well, Spain was a playground for French tourists. For most of this period, exchange rates and prices favored Spain over France as a vacation spot for those French who flee south toward the sun every summer. For those French staying on the Ruby Coast, even in towns and villages north of Cerbère, a short day trip over the border to Portbou is a simple affair. For those on the return trip on the coast to France from further south in Spain, Portbou is also an easy stop. In sum, while Cerbère is seen by those coming south as the exit from France, Portbou has the advantage of being both an entry and exit point into and out of Spain.

Intervillage Commerce

Although there are few natives of Cerbère fluent in Catalan, a good deal of cross-border contact occurs between Cerbère and Portbou. The vast majority of the shop owners in Portbou speak French. Before the adoption of the euro as the common European currency, French money was accepted in all the shops and restaurants in the village. The peseta, however, was never accepted in Cerbère. In fact, it was only in the middle of the 1990s (before the introduction of the common European currency) that the cash machine in Cerbère provided people a choice between Spanish and French currency. Some informants told us that the former mayor, and important local businessman, opposed the installation in the village of a cash machine equipped to give either Spanish or French currency. Among his many enterprises was the local money changing office!

Given the continuing disparity between prices in the two villages, people from Cerbère frequently shop in Portbou, particularly for food during the Friday market. On the other hand, since French salaries are superior to those in Spain, workers from Portbou frequently seek employment in Cerbère, often on the railroad and related services when jobs are available. Several informants, particularly young single individuals in Cerbère, told us that Portbou was more attractive than their own village for its social life. The former offers a greater choice of bars and cafés with significantly cheaper prices for food and drink.

Social Life

When comparing the two villages it is clear that the community-based social life in Portbou is considerably richer than in Cerbère. The annual *festes* are lengthy affairs and are attended by almost everyone in the village. These include picnics, musical entertainment, and a series of balls. In winter, Kings' Day (Twelfth Night) is a major event during which the three Kings (local people disguised in royal costumes) arrive from the sea by boat and mount horses from which they distribute candy to the children. The procession winds through the village until it reaches the community center where a crèche is constructed each holiday season. The resident priest in Portbou (there is no resident priest in Cerbère) runs a discussion group for married couples that meets once a month. These evening meetings end with a communal dinner in a local restaurant. We attended one of these meetings, including the dinner, and found the ambience to be warm and friendly. Though many of the inhabitants of Portbou join in the Sardana at festivals (the Sardana is the Catalan national dance and a strong symbol of Catalan identity) there is no formal dance group. (Such groups have been formed in Cerbère, however.) An important cultural activity is the village choral that has many dedicated participants and is expertly led. Also a small group of Portbou residents publish a local and, in my opinion, high quality magazine, *El Full* (The Leaf), appearing four times a year. Published completely in Catalan, it deals with local and national issues as well as culture and politics. Portbou boasted a night club, now closed, that was attended by the youth of both villages and may have been responsible, at least in part, for some of the intervillage marriages that occurred in the recent past. There is, in addition a dedicated theater group. Finally, many women in Cerbère prefer to have their hair done in Portbou although there is one hairdresser in their own village. There is a choice of three such facilities in Portbou. The most popular one is a favorite place for sharing gossip among the older women of both villages who speak Catalan.

Sports

Portbou boasts a set of popular junior football clubs supported by the majority of locals who attend games in the village stadium on the outskirts of town. The space is functional but rather run down and there is a plan to construct a more elaborate facility as soon as funds are available. (The project was still on hold during our last visit in 2005.) Cerbère has no football stadium but school age children often have the chance to play on teams of the neighboring villages, Banyuls and Port Vendres. On the other hand, it does have a large public gymnasium with an excellent basketball court. During our first visit to Cerbère in 1996, we witnessed a game played by the older girls of the village against another local team. It was enthusiastically attended by a large group of village team supporters. While this is the only major sports activity in Cerbère that attracts a relatively large number of spectators, tennis, para-sailing, and *boule* are popular sports on the participatory level. (*Boule* is a popular French bowling game in which teams compete with one another in an attempt to get their heavy metal balls as close as possible to a small round wooden target.) There is also a sailing club responsible for the upkeep of the floating pier, taken up every year at the end of the fall season. Members of this club, all men, also participate in an annual race among groups of typical Catalan boats with their lanteen rigging and phallic prows. In addition, there is an active divers group, a small private business venture that attracts participants mainly from outside of Cerbère.

Portbou also has is own *boule* club with a playing ground located on the other side of the tunnel, which allows one to pass under the railroad tracks. What is by far the largest social club in either of the villages, however, is in Portbou which, like many other Catalan villages and cities, boasts an active *penya* (football fan club) that functions primarily to support the Barcelona team. The team, affectionately known as "Barça," has a following throughout Catalunya that is close to cult-like. But the *penya* is much more than this. The membership cuts across all age groups and the men and women are of either Castilian or Catalan origin. It thus serves a major unifying social function in the village. While it has no public gymnasium or indoor basketball court, Portbou does have a small, well-equipped, private gymnasium facility, with a good selection of exercise machines. Members come from both villages. The owner is a dynamic young woman who teaches classes in gymnastics to youngsters of the village. She has had considerable success. The teams she coaches have won several Catalunya wide competitions.

Symbolic Links

While the French government has gone out of its way throughout history to suppress regional languages and cultures, certain purely symbolic ties between Catalans on both sides of the international border are tolerated as long as they have no overt political overtones. There are two *Sardana* groups in Cerbère (at this writing one appears to have been disbanded) and the nursery and elementary schools offer their students a certain degree of exposure to the Catalan language and culture. The Perpignan rugby team (USAP) supported by many Cerberians bears the Catalan colors on their uniforms and is seen as a special identifying mark of French Catalunya. Of greater importance are the personal intervillage links, as weak as they might be, that persist to this day among those few residents of Cerbère who look across the border to Portbou where Catalan culture still flourishes to a great extent. There is the sometimes grudging admiration many in Cerbère hold for the people of Portbou, particularly their sense of commerce and the *joie de vivre*, manifested in the café life of the village as well as in the numerous festivals celebrated during the year. There is a certain pride among some Cerberians for the existence of an autonomous region across the border in which the interests of a local culture are protected and promoted.

Beyond these social and economic manifestations of cross-border solidarity there exists in Cerbère a material reminder of the blood ties that united the people of the two villages during the First World War, a war in which Spain did not participate, but in which the youth of Portbou gave their lives fighting, if not for France, for their Catalan brothers and sisters. This is Cerbère's monument to the dead of that war that stands in a place of honor in the village and before which, every year on Armistice Day, people come to remember the sacrifices made by the youth of both villages. The sense of unity that is marked by this cenotaph was manifested once again in reverse and in more recent history when a group of young men from Cerbère, most, if not all, of Catalan origin, crossed into Spain to fight for the Republic against the Franco insurrection. Those few left in Cerbère who fought for the Republic look back on that time with pain because of the harshness of that war's bitter end, but also with a sense of pride and brotherhood.

An Important Symbol of Cross-Border Unity

Every June 23, Mediterranean people celebrate the coming of summer with the festival of St. Jean (St. John) that takes place a few days after the summer solstice. This festival involves the burning of an enormous bonfire (in Catalunya it is piled high with discarded furniture, old toys—things to get rid of from the past year—for the festival is one of renewal). During Franco's reign the practice was banned in Spain, as a danger to public order. The birth of a new meaning for this custom came about in the 1950s. An informant from Northern Catalunya involved in this transformation described it to us as follows:

> Can you tell us about the festival of St. Jean and what it means in Catalunya? Yes, once the festival was a tradition throughout Europe, but in recent years its celebration has slowly diminished in the region with the exception of France and Spain. There are still many customs and beliefs associated with it in those areas where it is celebrated. It is believed, for example, that when people jump over the flame they will have three years of good luck. If a person jumps over the fire with a child in hand that child will have a year of good health. The festival celebrates the coming of summer, the longest day of the year. What special political meaning does the festival have here? When the festival was forbidden by Franco in Spain, it continued to be celebrated here. In the 1950s seven or eight men from this region decided to carry the flame to the top of the Canigó mountain each year where the flame was to become the symbol of unity among all Catalans. After a few years the beginning of the ceremony moved to Perpignan in the castle of the Kings of Mallorca where the flame was kept alive throughout the year. From there it was taken to top of the Canigó. Meanwhile, hikers from many villages in Roussillon would each carry small amounts of wood to the mountain top to symbolically feed the flame that had already arrived. After that volunteers would secretly carry the flame southward into Catalunya. Franco, of course, forbade the clelebration and to counter his interdiction a human chain was formed to carry the flame southward in secret. The message for the south was one of courage and brotherhood. It symbolized the fact that although under Franco "night" reigned in the region, soon the light of day would return bringing liberty with it. When it arrived in a village people aware of its presence could not let on because the police were present to prevent the celebration. So, on arrival, the flame was transferred to a storm lantern to safely pass the night hidden from the authorities. In preparation for the next day every village had a place in which a certain amount of liquid fuel was stored. When the time came the flame was taken from the lamp and transferred to a small piece of corrugated cardboard, and, at the same the church bell would ring. As the bell sounded a Catalan flag would gently tumble down to the village plaza just below the church. Today it's difficult to put the meaning of Saint Jean in the context of the epoch and to keep the importance of the festival alive among the young. Here some young men carry the flame to villages in the region. The message is translated from Catalan into French, but now the custom is less militant and more folkloric. Sometimes, however, someone from here will carry the flame

all the way to Paris for the Catalans who live there. Some times it is even carried as far as the Catalan community in Quebec!

What follows is the message carried, along with the fire in 1993:

> *Flama del Canigó—Missatge de Sant Joan*
> The flame that you receive every year from the Canigó, and which is renewed each year, has a special significance.

We remember that those who carry the flame
wish to demonstrate, with this action, the unity of
all Catalans, and erase the borders between the two
sides of the Pyrénées.

This day we can say—as a fact—that our ancient
dream has been realized: we have broken the artificial
frontier, but we have not yet recuperated our national
rights.

The poet Miquel Duran wrote in 1916:
   . . . and above the spaces, from one side to the
other, the cinders kiss and mix together along with the
symbolic sparks of the fire.

Like them, we desire that this flame, ambassador
of the *Països Catalans* [Catalan-speaking regions] unites us once more,
and that this night of St. John is, for everyone, a night of good spirits,
of peace and hope, and a night of brotherhood with all the other
European people.

The festival of St. John (St. Joan in Catalan) is still celebrated in Portbou with the coming of summer. The custom of carrying the flame from France into the *Principat* remains the core of the event as French Catalan nationalists continue to take it from Perpignan to the Canigó and, from there, to different points across the border. While the people of Cerbère do light the fire on St. John's day in celebration of the summer solstice, they no longer participate in the circuit that carries the flame southward. Portbou receives its share of the flame from Collioure (a Northern Catalan town that has maintained the custom).

What the Two Villages Taught Us

From working, even for a short time, in Portbou and Cerbère it was obvious that important differences exist between the two villages under the influence of past and recent events. The most striking difference, of course, is the fact that Cerbère can no longer be considered Catalan in any real sense of the term, whereas Portbou, even though it is by no means typical of nonborder Catalan villages because of its large transient population primarily of Castilian origin, does remain Catalan by virtue of its legal and official linguistic identity. It is important, however, to take note of the fact that Catalan identity in Portbou is not a unified concept for all who identify themselves as Catalans. It ranges from strong feelings of nationalism among those who would like to see Catalunya an independent state to those who have a strong attachment to Catalan culture and language, but see Catalunya as an integral part of Spain. There are also inhabitants of the village, some of whom intend to stay beyond their working years, who identify strictly as Spaniards and not at all as Catalans.

All of Catalunya, not the least Portbou, has been hurt by uncontrolled immigration into its territory from other parts of Spain and Latin America. New arrivals often see no need to learn Catalan and some even express hostile attitudes toward it. Additionally, we have been told that there are some Catalans who feel that their language is in some sense inferior to Castilian. It is also the case that most, if not all, Catalans speak Castilian, whereas Castilian speakers need not learn Catalan in order to function economically and socially in Catalunya. A related problem for the language is the fact, already noted, that while all children in Catalunya are schooled in Catalan the tendency to speak Castilian in the school playground is widespread.

Another striking demographic fact is the overly large presence of retired people in both villages. This unbalanced age pyramid has a negative effect on the social and economic life and limits the pool of children available to the village schools. Most retirees are interested in finishing out their lives in peace and have withdrawn from active participation in the social and political life of their village.

The status of Portbou as a part of the autonomous region of Catalunya has had a positive effect on the village's attempt to cope with the changed economic conditions discussed earlier. The government in Barcelona cares about Portbou and its future. It has supported more than one important project to make up for the loss of business associated with the now open border. For the last several years Portbou's village council has been dominated by the same center-right Catalanist party, *Convergència i Unió*, that ruled the *Generalitat* from the restoration of autonomy after Franco's death until the fall of 2003. This political association has clearly kept ties between the village and the Catalan government strong. Additionally, the central Spanish government in Madrid, responsible for the national road leading from the south to Portbou, has agreed to reduce the isolation of the village by constructing an expensive tunnel between it and Colera. Portbou has also received substantial aid from the European Community. Many of our informants said that it was easier in Spain to get this type of aid than in France. Cerbère, on the other hand, gets very little aid from the highly centralized French government. Though Cerbère did get some economic support when the handicapped patients' hospital was constructed near the village, this facility does not provide much employment for villagers. The problem of attracting new young people to Cerbère has been compounded by the fact that the village council has been negligent in creating affordable housing for workers who might wish to live there. Many employees of the railroad live as far away as Perpignan commuting to work and shopping in Perpignan's large malls, thus depriving Cerbère's struggling retail shops of potential business.

As already noted, the fact that Cerbère remains an exit point from France and does not attract many long-term vacationers from either side of the border limits the role of tourism to the short summer season. This is due to the still existing economic differences between France and Spain. It is also due to a general lack of hotel space, although Cerbère is somewhat better supplied with rooms, at least during the high season, than Portbou. In contrast, Portbou continues to maintain its advantage over Cerbère as the entry point into Catalunya for French and some other European tourists. Thanks to the Walter Benjamin Monument a fair number of German tourists stop in the village to see it and pay their respects to the philosopher. The hotel situation and tight housing do not help although Portbou is more attractive during the vacation period than Cerbère since it has a plethora of restaurant-cafés and a series of major summer activities. The high point for the village is the festival of its saint's day that lasts a good part of one week. At that time the village is tightly packed with visitors. Portbou receives good marks every summer for its seawater quality as well as the cleanliness of the beachfront. Each summer the park by the sea is festooned with a flag awarded by the *Generalitat*'s health department for the high quality of its beaches. Tourism could easily be enhanced if the quality of restaurant food were to improve given the lower food and hotel prices in Spain perpetuated by differences in the added value tax between the two countries. While the French tax is an astronomical 21.6 percent it is only 7 percent in Spain. Cerbère's beaches are also clean but overall it is a very dirty village. Dog droppings are rarely removed either by dog owners (not a French custom yet) or by employees of the village council, and apart from the seafront, which has been spruced up of late, it has a somewhat seedy look. The Belvedere hotel in Cerbère, if funds for its restoration could be found, might turn into a valuable asset for tourism. It is a unique and rather startling example of Art Deco architecture in the region and, as has been mentioned, occupies a spectacular site over the railroad tracks.

Cooperation between the two villages would be a natural solution to part of the economic difficulties they both face. I have already noted the reticence of the French government to become involved in such projects. But there are also legal and cultural differences between France and Spain, as trivial as they might be, that get in the way of cooperation. I have already remarked on the difficulties faced by Cerbère in obtaining aid from the European Community due to conflicts with French law. On the cultural side, the people of Cerbère, most of whom have a very positive view of Portbou, point out that different holidays and very different eating times make cooperation difficult in arranging intervillage events that could focus on what's left of Catalan traditional culture.

In chapters three and four, I present interview material from both villages. In general anthropologists use such material, just as I have in this chapter, as the basis for their conclusions but rarely include long sections of them in the book. I prefer as much as possible to let the people speak for themselves because I believe that a culture is much more than a text, a concept often expressed by postmodern anthropologists. It is people as they live their lives, their desires, their failures, and their successes. In all cases I have transcribed and translated the tape recorded interviews myself from either French or Catalan and, then, after choosing which ones to include in this book, edited them for grammatical mistakes, excluded redundancies, cut sections that I believe to be inappropriate for the subject of this study, and excluded material that would make it easy for informants or other inhabitants of the two villages to identify those who were kind enough to grant us the time and patience that these interviews required of them.

Before I turn to these interviews, however, I offer a short review of nationalism in general followed by some crucial information concerning Spanish and Catalan history to show how modern Catalan nationalism fits into the scheme of nonviolent resistance to Castilian domination.

Chapter Two

Framing the Study: The Origin and Meaning of Nationalism

The sociologist, Ernest Gelner, offers useful definitions of the nation, the state, and nationalism in his book *Nations and Nationalism*. Aware that these three terms are often confused, he notes that the definition of nationalism is parasitic in relation to two other terms. Following the German sociologist, Max Weber, Gelner sees the state as that "agency within society that possesses the monopoly of legitimate violence" (3). He goes on to say that this definition applies to modern states very well but offers the caveat that in the medieval period, for example, feudal states did not fit the model since they tolerated violence in wars between competing fief-holders.

Gelner offers two definitions of nation neither of which he finds completely satisfactory. These are as follows:

1. Two men are of the same nation if and only if they share the same culture, where culture means a system of ideas and signs and associations and ways of behaving and communicating.
2. Two men are of the same nation if and only if they *recognize* each other as belonging to the same nation. In other words, *nations maketh* man; nations are the artifacts of men's convictions and loyalties and solidarities. (7)

For Gelner *nationalism* "is a political principle which holds that the political and the national unit should be congruent" (1).

From this it follows that nationalist *sentiment* "is the feeling of anger aroused by the violation of this principle, or the feeling of satisfaction aroused by its fulfillment." A nationalist *movement* "is one activated by a sentiment of this kind (1). So that nationalism "is a theory of political legitimacy, which requires that ethnic boundaries within a given state . . . should not cut across political ones, and, in particular, that ethnic boundaries within a given state . . . should not separate the power-holders from the rest" (1).

Gelner says that false consciousness is a negative symptom of nationalist ideology. "Its myths reverse reality: it claims to defend folk culture while in fact is forging high culture; it claims to project an old folk society while . . . helping to build up an anonymous society" (124).

He expresses an even more negative judgment when he says

> Nationalism tends to treat itself as a manifest and self-evident principle, . . . and violated only through some perverse blindness, when in fact it owes its plausibility and compelling nature only to a very special set of circumstances, which do indeed obtain now, but which were alien to most of humanity and history. It preaches and defends continuity, but owes everything to a decisive and unutterably profound break in human history. . . . Its self-image and its true nature are inversely related with an ironic neatness seldom equaled even by other successful ideologies. (125)

I turn now to what in my opinion is the most important contribution to our understanding of nationalism: Benedict Anderson's book, *Imagined Communities*. The title of his book reflects the central idea that the nation is a construct of human culture and that such entities we call nations are not coterminous with mankind (7). Furthermore, the nation is *imagined* as *sovereign*, hence his title, a change that occurred recently in history with the coming of the age of enlightenment and the revolutionary changes it engendered. It was these changes that destroyed the "legitimacy of the divinely-ordained hierarchal realm" (7).

Although, like Gelner, Anderson locates the birth of nationalism in relatively recent history, he offers a direct challenge to Gelner when he says

> The drawback to this formulation [the notion that nationalism is a fiction because it is an invention], however, is that Gelner is so anxious to show that nationalism masquerades under false pretences that he assimilates "invention" to "fabrication" and "falsity," rather then to "imagining" and creation. In this way he implies that "true" communities exist which can be advantageously juxtaposed to nations. In fact all communities larger than primordial villages of face-to-face contact (and perhaps even these) are imagined. Communities are not to be distinguished by their falsity/genuineness, but by the style in which they are imagined. (6)

According to Anderson the destruction of the divinely ordained realm is linked to the fall of Latin as the common language in much of

Europe that was replaced by a wide range of local tongues tied to their individual territories. While the former states, united by the common language, were loose conglomerates of people located within indistinct borders and controlled from the center, the new, emerging, entities took on both territorial and linguistic coherence.

Anderson goes on to say that this process led to new modes of defining the world that made it possible to visualize "the nation." This, in turn, was accompanied by the accepted notion of time.

> What [took] the place of the medieval conception of simultaneity along time ... is an idea of homogeneous empty time, in which simultaneity is, as it were, transverse cross-time, marked not by prefiguring and fulfillment, but by temporal coincidence, and measured by clock-calendar. (28)

This change in the conception of temporality is linked by Anderson to the invention of printing that in turn made possible the rise of the novel and the newspaper both of which provide for the kind of community imagined as the "nation."

> What, in a positive sense, made the new communities imaginable was a half-fortuitous, but explosive inter-relation between a system of production and productive relations (capitalism), a technology of communication (print) and the fatality of human linguistic diversity. . . . these print-languages laid the basis for national consciousness in three distinct ways. First and foremost, they created unified fields of exchange below Latin and above the spoken vernaculars [national versus regional languages]. . . . Second, print-capitalism gave a new fixity to language. . . . Third, print-capitalism created language-of-power different from the older administrative vernaculars. (42–45)

Anderson then goes on to cite one other important factor in the emergence of the nation. This is the fact that all the new state-nations founded through revolution in South and North America were formed by "Creoles" of common descent and language shared with those of the mother country who ruled them. Thus: "language was never even an issue in these early struggles for national liberation" (45). According to Anderson, such revolutions were spawned by the political exclusion of these Creoles who, regardless of their ancestry, were considered inferior to those who controlled the reins of power at home.

Back in the home country, the rise of the bourgeoisie was marked not by its unity of blood. Instead it "came into being only in so many replications. The factory owner in Lille was connected to the factory owner in Lyon only by reverberation" (77).

> The independence movements in the Americas became as soon as they were printed about [in the mother country] "concepts," "models," and indeed "blue prints".... Out of the American welter came these imagined realities: nation-states, republican institutions, common citizenship, popular sovereignty, national flags and anthems, etc., and the liquidation of their conceptual opposites: dynastic empires, monarchical institutions, absolutism, subjecthoods, inherited nobilities, serfdoms, ghettoes, and so forth.... By the second decade of the nineteenth century, if not earlier, a "model" of "the" independent national state was available for pirating. (81)

Finally Anderson offers another important point, particularly in the light of recent ethnic conflicts in Africa and the Balkans, when he distinguishes nationalism from racism. He says that

> nationalism thinks in terms of historical destinies, while racism dreams of eternal contaminations transmitted from the origins of time through an endless sequence of loathsome copulations: outside history.... The dreams of racism actually have their origin in ideologies of *class*, rather than of nation. (149)

Another author who deals at length with nationalism and, in particular, ethnonationalism is Walker Connor. Along the way Connor stresses the fact that many nation-states that profess to be unicultural are, in fact, made up of many indigenous ethnic groups. Connor puts the blame for this misinformation on theoreticians of nation building since they generally ignore ethnic diversity within the nation. Citing a total of 132 contemporary states he notes that only 12 are ethnically homogeneous. Furthermore, he remarks that in 31 of these states the largest ethnic group amounts to only 74 percent of the population. Astonishing as it might seem his statistics reveal that in 39 cases within the sample the largest ethnic group fails to account for half of the state's population.

Another important point made by Connor is his contention that scholars of national identity tend to equate nationalism with loyalty to the state rather than loyalty to the nation. The immediate and unsurprising result of this is the common confusion between the "nation" and the "state" that leads scholars to ignore what ethnonationalism actually is. This leads him further to claim that nationalism cannot be explained in materialist terms but is, rather, psychological in nature.

He says specifically, "Those who have most successfully mobilized nations have understood that at the core of ethnopsychology is the sense of shared blood, and they have not hesitated to appeal to it. When used pristinely *nation* refers to a group of people who *believe* they are ancestrally related. It is the largest grouping that shares such a belief" (212).

We shall see below that although negativism toward others (most strongly against Moroccan immigrants) exists among certain segments of the Catalan population, the official attitude toward outsiders in Catalunya is one of assimilation based on language and not on genetics. In fact I have met some Catalans who prefer to see immigrants coming to their region from Morocco to those from within other regions of Castilian-speaking Spain and Latin America. This preference is linked directly to language. Our informants who take this point of view do so precisely because of they believe that Arabic-speaking immigrants to Cataluña will learn Catalan more readily than those from different parts of Spain and other Spanish-speaking countries. It has been our experience that many Castilian speakers who have resided in Catalunya for many years either do not want, or do not bother, to learn Catalan.

A Look at the History of Spain in Relation to Central Versus Regional Power

In this brief summary of Spanish history I rely largely on an article, "The Roots of the National Question in Spain" by Simon Barton. The author looks initially at the geography of Spain pointing out that, at first glance one might expect it to be a unified entity since it sticks out from the rest of Europe as a peninsula bordering on the Mediterranean Sea to the east and the Atlantic Ocean to the north and north-west. To the north-east, it enjoys semi-isolation from France due to the Pyrénées. The author goes on to point out, however, that if one looks at the internal geography of the country one is immediately struck by the fact that it is bisected by several mountain chains that divide one region from another. Even the central region, the high plateau, the Meseta, is ringed on all sides by the mountain ranges of the Sierra da Estrela, the Sierra de Gredos, and the Sierra de Guandarrama. He goes on to say that though this rugged geography has in fact tended to encourage the development of regional and separatist movements, political, economic, and cultural factors are equally important in this respect.

Looking back at Spanish history the author notes that the exact date for the emergence of Spain as a nation is a matter of contention. The conventionally accepted event that is said to have brought about Spanish political, if not cultural unity, was the marriage of Isabel and Ferran (Ferdinand) in 1469 that united Castile and Aragon. It failed to produce a united country, however, since each partner in this marriage retained control over his/her respective territories. For this reason Barton suggests that in reality the probable date for the establishment of political control was the imposition of the *Nueva Planta* (New Plan, literally "new from the ground up") by Philip the Fifth, in 1716, a law that not only banished the use of Catalan in all the areas in which the language was spoken, but also closed all the universities in which Catalan was the dominant tongue. He notes, correctly, that even this cataclysmic change, while it did indeed bring about political unity, did little to create a national consciousness.

Nonetheless, not only was the marriage of Ferdinand and Isabel politically successful, it also led to the enrichment of the country as a result of the conquest of much of Latin America. The direct consequence of this was that, in a short period of time, Spain became a major European power. Its political fortunes on the European continent also grew successful with the coronation of Carlos the Fifth in 1516. The country's slide downward, however, began shortly afterwards and continued for three centuries. The final blow was perhaps the fall of the Spanish Empire in the New World that expired definitively with the American victory in the Spanish American War. The author says of this final blow to Imperial longings:

> In an atmosphere of anguish and self-criticism, a group of intellectuals ... set about examining the causes for Spain's plight and putting forward solutions for her salvation. It was not just the historical development of Spain that was being probed here, however, but the essence of the Spanish soul. ...
>
> Spain's political decline was mostly examined on a very superficial level with any number of facile explanations being offered to account for the "Spanish predicament." (108)

In the discussion that follows on these remarks Barton points out that the well-known Spanish philosopher, Ortega y Gasset, and other less well-known authors made the claim that Spain's decline was due to the lack of a strong aristocracy. He says that another famous writer and philosopher, Miguel Unamuno, went so far as to dismiss the unity of Spain as an absurd nationalist myth. He then moves on in his review of this controversy to another group of scholars who attacked Unamuno's point of view with the response that Spain's emergence as a nation "had been the triumphant product of the continuous unifying impulse that had driven their ancestors.... For [these] nationalist historians, the weakness of the ties which bound Spaniards together was to be attributed above all to their unity of Character" (109). This weakness was diagnosed as the innate individualism of the Spaniard [and] "not any fundamental ethnological or linguistic differences between, say Catalans and Castilians, that explained the strong regional feeling that existed in Spain" (109–110). This very interpretation was taken during the Franco period in which the dictator promised a rebirth of patriotism and national solidarity.

After reviewing and criticizing the different renderings of Spanish history through the ages, Barton concludes that Spain's inability to maintain unity occurred because no important attempt was made to foster a sense of Spanish identity among all the inhabitants of the peninsula. Additionally, regional discontent with the center was compounded by the latter's economic decline just as the regions developed their local economies. "Thereafter, as the economic initiative passed to the regions, ... political power remained firmly rooted in Madrid. There developed a widespread conviction that the center was running Spain at the expense of the periphery, a feeling that has persisted to this day" (121). This is indeed the feeling, at the present time, of those Catalans who continue to press for greater autonomy, or even independence from Spain. Among the data that confirm this claim are the following statistics published by the *Generalitat* of Catalunya at the end of the 1990s pointing the finger at what it considers to be economic discrimination against Catalunya in the redistribution of Spanish national tax revenues: in all of Spain the redistribution of funds to the regions was one to 1.5 percent of federal taxes while Catalunya received only 0.75 percent; the Spanish state invested 556 euros per inhabitant for the region of Madrid while Catalunya received only 329 euros per habitant, representing 41 percent less. Madrid had 9 percent of all the toll free autoroutes and 0.75 percent of toll roads while Catalunya had 19 percent of the free autoroutes in all of Spain and 42 percent of the toll roads.

Josep-Lluís Carod-Rovira, president of *Esquerra Republicana de Catalunya* (the left-independence party and then member of the Catalan Parliament) published an article, *La Gran Mentida* (The Big Lie) in the September 15, 2004, issue of the Catalan language newspaper,

Avui. What Carod calls the "Big Lie" is the persistent claim that Catalunya benefits from unfair fiscal advantages under Spain's tax laws. Carod argues, to the contrary, that Catalunya has continuously suffered from fiscal discrimination on the part of the central Spanish government. In stating his case he points out the following:

> There are regions [outside of Catalunya] in which every elementary school student has his or her up-to-date computer provided free of charge by the State along with all necessary textbooks. The difference between the cost of a regular driving license and a professional truck license is waved in much of Spain, but not in Catalunya. Free burials are provided in all regions of Spain in comparison to Catalunya. Better financed health coverage is provided to all the Spanish autonomies with the exception of Catalunya. In spite of the fact that 17.5 percent of the Catalan population lives in poverty, the region has never received its fair share of financial support for poor families from the National Government. Students in Catalunya are forced to take classes in run-down prefabricated buildings. Catalunya suffers from an antiquated infrastructure, a health system in severe debt, and businesses that grow only on the basis of their owners' entrepreneurial spirit without government subventions. All of this has nothing to do with national solidarity. . . . Will the day come when others behave in solidarity with us!

Some Important Dates in Catalan History

What follows are selected dates in Catalan history that illustrate the depth of the Catalan claim to national identity, independent or partially independent of Spain. Additionally this section deals with the development in the long period of the Catalan-Aragon federation, of innovative and, for their times, democratic institutions. The information is drawn from, among other sources, a summary of Catalan history provided by *Omnium Cultural*, an association for the promotion of Catalan culture.

The first ruler of an independent Catalunya was Guifré the Hairy who in 878 established the oldest line of descent in medieval Europe. The name "Catalunya" appears for the first time in 1114 in a work known as the *Liber Maiolichinus* and in the first known document in the Catalan language, the *Liber Ludiciorum*, dated in the year 1150.

By the end of the tenth century the descendants of Guifré formed a patrimonial condominium over the subregions of the Cerdanya, Besalú,

Barcelona, l'Urgel, Girona, and Osona. Under this regime, the territory was independent of the Frankish kings and in 1258, the Catalan King, Jaume I, and Louis II of France signed an agreement in which the French gave up all claims to Catalan territory south of the Pyrénées and Jaume renounced his rights to all land north of the same mountain chain.

In the year 1150, Catalunya developed a feudal constitution well before the Magna Carta was passed in England. This document recognized equality between burghers and nobles. In his book *Catalonia, Nation Building Without a State*, Kenneth McRoberts says of this historic development:

> By many accounts, the system of covenants and law implanted a world view, dubbed *pactism*, that has become an enduring trait of Catalan society. Under normal conditions, it is argued, the legal and political life of Catalonia has been governed by *pactist* assumptions: rules emerge from contracts that parties make of their own accord, and social life in general should be based upon bargaining and accommodation rather than on force or domination. (10)

During the Middle Ages, Catalunya was the only Hispanic territory to expand outward. The Catalan-Aragon alliance extended its territory throughout the entire western Mediterranean, including València, Mallorca, Sardinia, and Corsica, as well as Athens and Naples. In the fifteenth century, the "Corts of Barcelona" (a Catalan legislative body) established the principle of a "limited" monarchy that was obliged to govern according to laws, while guarding a degree of royal power.

In 1659 at the end of the war of the Segadors (a peasant uprising against France and the central Spanish government) the Catalan territories north of the Pyrénées were ceded to France without any consultation with the Catalan rulers. After taking over this territory the French government abolished all Catalan institutions within the former Catalan lands.

As already noted, Catalunya lost the independent control of its home territory, which it had maintained even after the marriage of Ferdinand and Isabel and the "uniting" of all the territories on the Iberian peninsula, when Philip the Fifth imposed the *Nueva Planta* (*Nova Planta* in Catalan) in 1716. During this period Castilian was imposed as the official language of the territory and was the only tongue taught in schools. The Catholic Church was also obliged to use Castilian in the teaching of religion. Although Catalan continued to be spoken in rural areas, it was not until the nineteenth century that

Catalan emerged once again as the language of the educated urban classes. This period is known as the "Catalan Renaissance."

In 1859 the medieval competition known as the *Jocs Florals* (Flower Jousts)—poetry contests in which the first prize consisted of three different flowers, one of which was a red rose—were reintroduced as a signifying the rebirth of the Catalan language as a literary form. The Institute of Catalan Studies, composed of Catalan scholars and intellectuals, was founded in Barcelona in 1907. Its work included the creation of an official set of linguistic rules leading to the standardization of the language. The first authoritative Catalan dictionary, authored by the polymath, Pompeu Fabra, appeared in 1932. By the end of the nineteenth century the first well-founded theories of federalism and nationalism were proposed by such intellectual figures as: Francesc Carreras i Margall, Valenti Almirall, Enric Prat de la Riba, Narcís Roca i Ferreras, and Domenec Marti i Julià. The earliest of these intellectuals were, for the most part, on the right of the political spectrum, but a leftist wing of Catalanism formed shortly afterwards that was particularly active in the political struggles before and during the Spanish Civil War.

In 1869, the "Pact of Tortosa" was signed by representatives of the former Crown of Aragon to create a confederation of its former autonomous territories.

Catalunya was one of the few areas in which Europeanism was popular among political activists and, to this day, Catalans tend to look north rather than west toward Madrid. Over the years, Catalan publishers have produced myriad translations from a wide range of European languages, including leading works of fiction as well as philosophy. For example, Catalan was one of the first languages into which the works of Nietzsche and the Norwegian playwright, Ibsen, were translated. Industry also developed rapidly toward the end of the nineteenth century, particularly the production of textiles. A considerable number of ambitious Catalans migrated to Cuba where many became rich. They returned to Catalunya with their newfound riches to become willing clients of leading architects, among these, Gaudí. Gaudí is known not only for his many "Art Nouveau" buildings (in Catalan *Modernista*), but also as the architect of the park Güell, built close to the Tibidabo mountain, at the western edge of Barcelona. Another important architect, Domènech i Montaner, designed many private residences in Barcelona, but is probably best known for his *Palau de la Música*, a unique gem, now classed as an international treasure by UNESCO, as well as the medical complex, the *Hospital de Sant Pau*.

In 1914, the *Mancomunitat de Catalunya* was established. This new political entity was responsible for coordinating the internal affairs of the four "*diputacions*" (Lleida, Girona, Barcelona, and Tarragona) into which Catalunya had been (and still is) divided by Spanish law in order to better control Catalan politics. The *Mancomunitat* was the first attempt by Catalan nationalists to abolish this division of Catalunya into separate political districts.

In 1923, the dictator, Primo de Rivera, assumed power and imposed repressive measures, including the suppression of the *Mancomunitat* and the Catalan language in favor of Castilian. The dictatorship lasted until 1930. Soon after the fall of Primo de Rivera, in 1932, the government of Catalunya (the *Generalitat*) made the region an autonomous part of the Spanish Republic. In 1936, the Civil War between the fascist troops of General Franco broke out ending in victory for the right-wing Spanish nationalists in 1939. Once again repressive action was taken against Catalan culture and language. In 1940, Lluís Companys, who had been the president of the *Generalitat* under the Republic and who, at the end of the Civil War, took refuge in France, was returned to Barcelona by the French and executed by the fascists.

Chronology of Events at the End of the Franco Period

In 1971 the clandestine *Assemblea de Catalunya* was established in Barcelona, composed of both legal and illegal nongovernmental regional and professional organizations. Its charter was to organize a nonviolent approach to democracy in the coming years. In the same year the new archbishop of Madrid, president of the Spanish Episcopal Conference, led the struggle for political democracy within the church.

In 1974, a new political party, *Convergència Democràtica de Catalunya* was founded. From 1980 until 2004 it was to be the major governing party in Catalunya under President Jordi Pujol. Although at times forced to form alliances with parties either of the left or of the right, Pujol served as president of Catalunya from 1977 until 2004, when the Socialist Party of Catalunya (the PSC) took power in a coalition with *Esquerra Republicana* and the *Initiativa de Catalunya—Verds*.

After Franco died in 1975 Juan Carlos the Bourbon was named constitutional monarch of Spain. Soon afterwards democracy was reestablished in Spain. The three "historical" autonomies, the Basque country, Galicia, and Catalunya were granted special status under the new constitution while the other regions of Spain took on the status of autonomous units within the state but with more limited power than that granted to the "historical" autonomies. Although Castilian was declared as the language of the entire country, Basque, Galician, and Catalan were recognized as co-official languages with Castilian. Soon afterwards, a language law was promulgated in Catalunya making Catalan the first language of the public school system. A clause in the new Spanish constitution, however, made it illegal for autonomous units to form alliances for economic and/or cultural ends thus blocking the historically Catalan-speaking provinces of València and the Balearic Islands from joining together with the *Principat* to create a larger political unit within the state.

In 1977, the first free elections were organized and democracy was reestablished in Spain. On the Catalan National holiday, September 11 of the same year, a massive and peaceful demonstration with over a million participants took place in Barcelona. Its major demands were the reestablishment of democracy throughout Spain and political autonomy for Catalunya.

In 1979 the statute of autonomy for Catalunya was approved by referendum.

The year 1983 saw the establishment of the Catalan public television station, TV 3, which broke the monopoly of Spanish state television. A second Catalan television station, Canal 33, was established in 1988. The year 1983 also saw the passing of the first law aimed at normalizing the use of Catalan in the *Principat*. The Catalan language was declared as the *llengua pròpia* (the official language of Catalunya). The new language law dealt with the place of Catalan in public life. Among the specific topics were the following: individuals were to be permitted to use Catalan, if they so chose, in the adjudication of court cases; Catalan was declared to be the official language of the administration; all laws passed by the Catalan legislature had to be published in Catalan as well as in Castilian. Furthermore, in dubious cases the Catalan version was declared to be the official document. (This section was declared null and void by the Spanish Constitutional Court.) Catalan was established as the language of education at all levels up to university studies with Castilian as well as Catalan to be taught in elementary and secondary schools; within the university, students had the right to be taught in the language of their choice (Castilian or Catalan) but all universities were required to teach Catalan to both teachers and students who did not speak it.

Additionally, the language law of 1983 made an exception to Catalan as the official language of the *Principat* for the Aran valley in the Catalan

Pyrénées. In this area, the local dialect of Occitan, Aranès, was recognized as the official language. To this day, of all the Occitan territories, the Aran Valley is the only place in which that language is officially recognized.

In 1998 the original language law of 1983 was strengthened for many areas of public life including the print media as well as film. The preamble to this new law states: "The object of this law is to further develop article three of the Statute of Autonomy of Catalunya to protect, encourage and further normalize the use of Catalan in all public areas of Catalan life with the exception of the Valley of Aran in which Aranese is recognized as the official language, and to guarantee the normal use of both Castilian and Catalan in the rest of the *Principat*."

Barcelona was declared to be the city of choice for the Olympic Games of 1992. Much of the city was renovated for this event and both Barcelona and the rest of Catalunya took advantage of the games to advertise Catalunya to the rest of the world. The major slogan of the games was that the Olympics were to take place in a country called Catalunya. Advertisements for the games included an upside-down map of Southern Europe with the *Principat* clearly demarcated.

In 1995, the *Infopista*, The Catalan Highway, became the official Catalan site on the Internet. Later the site's name was changed to *Vilaweb*.

In the year 2000, the Popular Party (of the right) won an absolute majority in the Spanish Congress. The two previous governments of the right and the left had depended on alliances with parties from the autonomies to govern. The new government under Aznar, no longer dependent upon the CiU (the Catalan center-right coalition), began a long attempt to weaken the power of the Basque and Catalan autonomies in the areas of language propagation and education.

In 2004, after the terrorist attack on the Atocha railroad station in Madrid, the Socialist Party of Spain (the PSOE) won the national election and began to govern in consort with regional parties.

Nationalism and Catalunya

This section deals with nationalism as it applies to Catalunya, reviewing in turn a set of journal articles that examine different aspects of *Catalanism* as it is, or has been, manifested in the *Principat*. I begin the discussion by looking at a well-known *Catalinist's* approach to the problematic, because of the often pejorative, status of the term "nationalist" in political discourse, particularly as it applies to Catalunya. The article in question bears the title "The Role of Nations in the European Construction" by Salvador Cardús i Ros, professor of sociology at the Autonomous University of Barcelona and a frequent contributor to the Catalan language newspaper, *Avui*. It was published in Catalan in a volume entitled *Els Nacionalismes factor de Violència* (I lack the complete citation for this work). The extracts below, however, are taken from an English copy of the article supplied by the author in the spring of 1997, and for which I also have no date.

Cardús i Ros begins with a general discussion concerning the negative atmosphere surrounding the term "nationalism."

> Making it a scapegoat for European disenchantment, systematically ignoring counter examples set by cases of nationalist-rooted peaceful self-determination from the separation of the Czechs and Slovaks to German reunification and concentrating all reflection on the almost unique case of the particular conflictivity of the former Yugoslavia. (10)

For Cardús i Ros, the problem may be reduced to the general conflation of the term "nationalism" with "totalitarianism" as an atavistic return to primitive social organization. This generally negative view leads some Catalans to feel uneasy about the concept. For this reason Cardús i Ros makes a distinction between the "voluntarist" nationalism that he characterizes as "rational and democratic" and its undemocratic form. He also distinguishes between the nationalism of internationally recognized nation-states, on the one hand, and the nationalism of people "who demand the right to self-determination on the other." This point is particularly germane in the case of Spain. Cardús i Ros says that the failure to see a nation-state as nationalist [as was so clearly the case under Franco and "which continues today, particularly among right-wing Spanish political figures] is because "nation-state nationalism" is commonly presented as "natural" while the same term used in relation to "nations" without states is considered to be artificial.

This leads Cardús i Ros to note the distinction so often made between nationalism and patriotism and, in this light, to quote the French author, Gilles Martinet, as saying that nationalism is an offence against reason. From this point of view patriotism reflects a love of one's own, while nationalism is based on hatred, contempt, and the fear of others.

Cardús i Ros sums up the virtues of a peaceful democratic nationalism as follows:

> A political entity culturally and politically stable and homogeneous enough to be able to integrate the diversity of the open world of today,

without creating marginal ethnic or cultural ghettoes. . . . It is national awareness that guarantees this necessary minimum level of collective identification which in turn makes the political community possible.

Nationalism becomes a condition for Internationalism.

Nation and nationalism are a guarantee of free participation in the global village. . . . [W]ithout the contrast of solid local identity, the global community is no more than a euphemistic way of referring to the provincialization of the modern world around a metropolis concentrating the greater part of cultural power in the large consumer and audio-visual industry. (18)

Thus, he concludes, "A truly united Europe will be that of the peoples and the nations or it will never be" (20).

The Rebirth of Catalan Nationalism

Modern-day Catalan nationalism or, as some prefer to call it, *Catalanism*, fits perfectly into the model presented by Benedict Anderson. Although Catalan identity is a fact of antiquity, the beginning of national sentiment came to birth only in the middle of the nineteenth century. The first Catalanists were from the growing middle class, excluding in most cases the very rich. In stricter terms Catalans were born in the professional class and among intellectuals, including artists and writers. This, perhaps, is why one of the earliest symbolic signs of this development was the restoration of the medieval poetry contest, the *Jocs Florals*, an exemplary form of symbolic nonviolence.

Catalan nationalism in its different incarnations found its followers almost from the beginning, on both the left and the right. Its form varied, of course, depending on which side of the political spectrum one stood. As it developed, it took on a set of symbols particular to the large, dispersed rural population with its deep roots in the countryside, as well as to a rising group of urban intellectuals. An interesting paper that deals with this process is by the anthropologist, Susan M. DiGiacomo.

The title of the paper, La Caseta i l'Hortet, refers to the small, rural house of the peasant farmer, the "*caseta*," and the kitchen garden of the small-scale farm, the *hortet*. In this work the author focuses on the relationship of capital and hinterland in Catalunya that the author describes as "having always been a complex amalgam of mutual attraction, contempt, envy, resentment and idealization. . . . characteristic of urban-rural relations in all Mediterranean societies but in each it takes a particular form" (160). These key symbols, she suggests, represent the concretization of a major source of "*seny*" (good sense). (Although DiGiacomo does not mention it, *seny* is usually coupled with another term of self-identification by Catalans. This is *rauxa*, which means impulse or emotionality.)

In much of Spain beyond Catalunya when asked to characterize Catalans the immediate reply is likely to refer to the Catalan's devotion to hard work and frugality, which can be expressed, depending on the prejudices of the informant in a positive or a negative light. This frugality includes "the logic of the slow accumulation of the typically, modestly sized but economically sound agricultural family embodied in the *casa pairal* (the family house and center of social and economic life) which served as the model for the industrial family enterprise, the *casa industrial*" (160). In this way the family in Catalunya has become the symbol of the nation.

At the end of the nineteenth century and the beginning of the twentieth, the mood shifted in the rapidly industrializing Catalunya from peasant nostalgia to progressive capitalization, particularly in the cloth industry. DiGiacomo points out, however, that the new "industrial landscape" was incorporated into the old vision as the anti-industrial movement, *Modernisme*, rapidly faded into the past to be replaced with a new forward looking philosophy, that of *Noucentisme* (Toward a New Century: The 1900s). The new ideal was to "urbanize" the countryside and, at the same time, "ruralize" the city. This result she says would be to realize a conception of nationhood and citizenship [that a spokesman for the new ideology, Antoni Rovira I Virgili, called "*La Catalunya-Ciutat*"] (161).

An important figure of this movement, and Catalan history, was Francesc Macià who was called upon by local political leaders, from a group known as *Solidaritat Catalana*, to run for election as their candidate. Macià became an independentist and, in 1922, a founder of a party known as *Estat Català* that was immediately outlawed. On the resignation of the dictator, Primo de Rivera, Macià, who had been in exile during the dictatorship, returned to Barcelona to help found the party know as *Esquerra Republicana de Catalunya*. He ran for election under its banner and, in 1931, was elected overwhelmingly.

> Macià was the "grandfather" to the "national family" because workers, peasants, and the middle classes all claimed him, all felt represented by him.
>
> A phrase that reappeared frequently in his speeches—"La casa i l'hortet"—has since become enshrined in popular memory, half

mockingly and half respectfully, as the encapsulation of Macià's social ideals (162).

DiGiacomo points out that rural Catalunya is, to this day, primarily a place of midsized holdings "cultivated either by peasant owners or by tenants on such long leases that they amount to perpetual use rights" (163). When the left-wing party, Esquerra Republicana (ERC), was restored to legality after the death of Franco, it took the image of *casa i hortet* and coupled it to the slogan "Enough talk and promises. Let's get to the heart of the matter."

In an article, "Catalan Identities," Ignasi Terradas, Professor at the Autonomous University of Barcelona, offers three types of identity expression and related behavior that he associates with three different geographical-social areas of Catalunya: the Pyrenean mountain region, the central highlands, and the large urban communities, such as Barcelona, Lleida, Tarragona, and Girona. The first of these identities is one that he says is lived rather than thought. This identity is the one that best fits traditionalist and romantic views of ethnicity and nationalism on the European continent. The second identity type, the one that is formally declared, is ethnic or national in character, and is expressed in economic and political propositions. The third type also has political and economic content expressed openly, and it is based on the conventional orders of the society in question. Terradas cautions, however, that the three types are permeable and interact with one another as well as with other types of identity. According to Terradas,

> Catalunya has produced an example of an extraordinary mixture of identities: interactions with the Spanish State and other peoples of Spain, appeals to traditional and modern ways of life, secular and religious policies, economic and cultural elites, etc. This shows ambiguity, contrast and contradiction both inside and outside [Catalunya]. (40)

Terradas clearly identifies language as the primordial, natural form of identity, a fact that appears in my mind to fit with the basic form of Catalan identity in general and which is investigated in this book. But, since language for this author is a *given* part of *Catalanitat*, he associates it with the first type of identity of which he says; "As such to the actor it cannot be considered as an identity because it is not used as such in his or her society"(40). Unless Terradas means to limit language to this first category exclusively within the context of nation-states in which a national language is taken for granted, I cannot agree with such a placement. Our data show that the vast majority of informants who identified themselves as Catalans, including those who are not Catalan nationalists, understood that the use and preservation of the language were essential to the continuation of the culture. Thus we would place the Catalan language within Terradas' *second* category, which he himself does later in the article. Additionally, the fact that the stated policy of the Catalan government's integration policy clearly stresses an intimate association between speaking Catalan and being Catalan demonstrates that it is clearly a conscious part of language politics in the *Generalitat*. Even when he finally does place language in the second category, Terradas weakens this affirmation, at least for Catalans, by adding the caveat that speaking Catalan is *not* sufficient to be considered a Catalan. Indeed, it is not sufficient, *in the abstract*, but speaking the language actively usually means an affiliation of some kind with the culture.

Elaborating on the second type of identity, Terradas says it is based not on psychologisms, but rather is a contemporary presentation of concrete activities and ideologies involving such areas as law, institutions, corporations, economic activities, and politics. Of this form he says, "In Catalunya this second type of identity is highly coincident with historical and geographical definitions of the nation" (43).

After offering a historical outline of Catalan relations with successive governments of Spain, which need not be considered here, he moves on to the period following the return to democracy after the death of Franco and offers an important judgment concerning the more recent development of his third type of identity. He links this shift in Catalan nationalism to the integration of Spain into Europe, the established boundaries of the Catalan-speaking areas, the integration of certain Catalan politicians into Spanish national politics, and the limits placed on cultural rather than class on the formation of national identity. He says this of the overall process:

> Individualism, crisis capitalism, elitism, integration into Spanish and European crisis management, and new forms of social-Darwinism are displacing the old ethnic and corporate formulations of national identity.... This type of identity, based on civic, educational, and democratic motives has a world model: the United States. (47)

Terradas warns, however, that this new, third type of nationality, most evident in the larger urban areas, will fail to achieve complete fulfillment since, as he says; "The identity proposed needs a state for its formal definition and ... without the state it is only a partial formulation of the third type of identity" (9). This danger is also perceived by many contemporary Catalan nationalists and is, I believe, justified by the facts of recent politics in which the right-wing Popular Party made serious attempts to dilute the importance of both the Catalan and Basque languages as well as the teaching of local history in the respective school programs of Catalunya and the Basque country.

"Catalan National Identity, The Dialects of Past and Present" by Josep R. Hobera, describes national ideologies as the dialectical interaction of the past with the present. Hobera points out that "though they project an image of continuity, they are pierced by discontinuities; though they conjure up the idea of an immutable ideological core and an adaptable periphery, in fact both core and periphery are constantly redefined" (11). The author also criticizes the received wisdom that the nation-state is the normal state of affairs in the modern world. He claims that it is more fiction than reality. He notes, although in stronger terms, the point made by Walker Connor cited above, that the nation-state is a qualified failure of European history since most so-called European nation-states are, in fact, multinational. He outlines an alternate approach in which he sees society as an evolving self-forming whole.

Agreeing with most students of Catalan history and, I might add, one of the main points made by Benedict Anderson, Hobera says that the modern form of Catalan identity began in the latter half of the nineteenth century, the period of rapid industrialization. He sees Catalan nationalism in this period as the result of five factors:

1. A strong ethnonational potential [by which I assume he means a long history of ethnic and sometimes political continuity].
2. The appearance of the Romantic nationalism characteristic of the revival of Catalan literary culture in the period known as the *Renaixença*.
3. The development of bourgeois society in the face of...
4. a weak and inefficient central state in Madrid.
5. A strong national church in Catalunya [which I would link to the growth of conservative Catalan nationalism].

Shifting to the effects of the Spanish Civil War of the 1930s, Hobera points out that this event was marked by strongly divergent conceptions of the nation and of the state. The Republicans, although they rejected Catalan independence out of hand, did hold to a version of autonomy for Catalunya, the Basque country and Galicia. The (Spanish) nationalists, as their name implies, propounded a strong centralized state based on fascist ideology.

The history of the forty years of Francoism in Catalunya is first and foremost the history of the survival of diffuse and contracted Catalan identity and then, in the 1950s, of the progressive consolidation of such identity, fundamentally at the linguistic levels, but also with a political dimension. (17)

Hobera uses this combination of factors to make what for him is a clear distinction between the nation and the state. On the one hand, he places the nation, which reflects such emotionally (affectively) loaded features as culture, freedom, romanticism, religious sentiment, and individualism. On the other hand, he characterizes the state, in opposition to the nation, as political rather than cultural, uniform rather than varied, compulsive rather than free, artificial rather than organic, rational rather than romantic, secular rather than religious, collective rather than individualistic, and instrumental rather than affective. He uses this schema to explain differences between the large immigrant population that began to arrive in Catalunya after Franco's victory, and that continues today, and the indigenous Catalans. To the latter the most common characteristic of these immigrants is that they come from the working class and do not, at least on arrival, speak Catalan. This, he says, creates a potential situation for conflict. Furthermore, although the majority of these immigrants are at the bottom of the social scale, they carry with them the official language of the state and can be seen as linguistic oppressors. Hobera offers the conclusion that because these two communities are there to stay the future depends on whether or not they can live together:

But the viability of Catalunya as a nation will depend, in the final instance, on the ability to develop a cultural and political identity that, while appearing to preserve the essence of the past, will allow the possibility of a common future for all those who live in Catalunya. (22)

It seems to me that the core notion of this hope for Catalunya lies in the future of the Catalan language. The *Generalitat* has wisely recognized this fact from the beginning of its restoration after the death of Franco by instituting programs in the elementary and high school systems, as well as in the business world, to make Catalan the primary language of public life. The first law to maintain the force of Catalan was passed in 1978 and was reinforced and revised in 1997. How successful these efforts will be, in the long run, remain to be seen. One can say, however, that the stress on language is a virtue to be maintained since in Catalunya, to its credit, it is culture rather than any putative genetic relationship to an often fictitious Catalan past that makes one Catalan.

In the light of these remarks I believe it is germane to end this section by citing an article by Miquel Strubell i Trueta, "Language and Identity in Catalunya," which deals with the relationship between the national identity of immigrants of Spanish origin resident in the *Principat* and their knowledge of Catalan. Of this relationship he says:

> Immigrants' identification with Catalunya is correlated more with length of residence than age at arrival, whereas the ability to learn and speak Catalan are more closely related to age on arrival. Eleven-year-old children of immigrants that use Catalan at home describe themselves as Catalan in most cases, as do those that speak Catalan well or very well. Those that know the least Catalan, on the other hand, tend in most cases to identify themselves as Spaniards. (91)

The latest statistics I have concerning the success of the language program in the *Principat* come via an article in a May 2004 issue of the Catalan language newspaper, *Avui*. This article makes a reference to the most recent study of the use of Catalan within households. The figures show that at present 51 percent of the inhabitants of Catalunya speak Catalan at home as their first language. Whether or not this is good news may be a matter of that old cliché: The glass is either half empty or half full.

Symbolic Forms Other than Language

Along with a strong emphasis on the standardized language, symbolic activities exist that support *Catalanism*. These have developed over time to sustain and encourage Catalan culture as well as to differentiate it from the rest of Spanish culture.

1. The Segadors

Let me begin with the Catalan national anthem, the *Segadors* (The Reapers). This song (actually written during the Catalan Renaissance of the late nineteenth century) celebrates a peasant war that began in 1640 (lasting until 1652) when the count of Olivares made the following recommendations to the King, Philip the Fourth: to create a military draft in Catalunya, to impose various taxes, to abolish certain parts of regional laws that up to then were respected by the state, and, finally, to impose Castilian law on these regions. These changes were to be put in place to support a war between Spain and the Hapsburgs that did not directly concern Catalans. The immediate cause of the War of the *Segadors* was a blood bath following the death of a peasant, in which the masses revolted and burned a number of public buildings and killed the *Viceroi*.

This violent reaction against the state was a defensive response to what were considered to be injustices and insults perpetrated by the central government against the autonomy. It is certainly not an example of nonviolent action by Catalans. It must be made clear, therefore, that Catalans were never nonviolent in the absolute sense. As I have already noted, in the Middle Ages they had fought wars of colonialism that spread their influence as far as Sardinia, and even Athens came under their sway for a time, but in the modern period beginning in the nineteenth century, Catalans restricted violent actions to defensive measure against outside dominance. It is well known, for example, that the Catalans (although there were some pro-Franco people among them) fought mightily along with their Basque allies and other Republicans against the Franco nationalists during the Spanish Civil War between 1936 and 1939.

The Catalan national anthem, The *Segadors*, thus celebrates a war fought to protect the integrity of Catalunya. Interestingly, when in the 1960s the future Catalan President, Jordi Pujol, led an audience in the Barcelona music palace (*The Palau de la Música*) in singing the *Segadors* as an act of defiance against the Franco government, he was put in prison for this offence against the Spanish state. Clearly the *Segadors* stands not as a call to war but rather as a symbolic expression of Catalan resistance to outside domination.

2. The Sardana and the Cobla (The Catalan National Dance and its Folk Orchestra)

The Sardana is, in one way, an invented tradition in that it originated in the north of the *Principat* long before it became a symbol of national resistance to the Spanish government. The Sardana is a circle dance performed by any number of participants to the accompaniment of the traditional Catalan folk orchestra, the *Cobla*, made up of various rather nasal double reed instruments, perhaps ancestors of the modern oboe, as well as brass, and a small flute played by a musician who also taps out the rhythm on a miniature drum. Although the Sardana is a circle dance, each performer remains in place within the circle. There are two rules for dancers who perform it. No one can be refused entry into the group and no one is allowed to join the dance between members of a couple (man and woman) who are already in the circle. The Sardana looks simple, but there is a complicating factor. At various points the music suddenly changes rhythm, speeding up considerably or slowing back down again. When this happens the dancers must adjust their steps and hop. To anticipate the sudden changes, at least one of the dancers must maintain an exact count of the measures in the music as he or she dances it, not an easy task especially as there are many different Sardana compositions with different counting tasks.

During much of the Franco period the Sardana was forbidden by the national government. This proscription was relaxed later before the end of fascist rule and the Sardana became a widespread means of displaying Catalan identity. While in Barcelona for language training in Catalan we were present at a Sardana performance during a celebration of one of the two patron saints of the city in which the entire Catalan governmental section of Barcelona was surrounded by an immense circle of dancers. This mark of Catalan identity is celebrated every weekend in most Catalan towns and cities. In Barcelona, the dance takes place every Sunday in front of the Cathedral.

3. *Castellers*

The youth of Catalunya enjoy another expression of national identity, the *Castellers*. This involves the very acrobatic, but perfectly organized, construction of a human tower that can reach up to ten levels. This feat is performed, in general, by nonprofessional youths. Except for massive support at the tower's foundation each level varies between one and five individuals who support the next tier. The larger the tower and the larger the number of individuals making up every level but the top, the heavier the burden supported from below. It is for this reason that the tower is always topped off by a small (and very brave) child, either male or female who, upon achieving the summit, rapidly descends sliding down the backs of the other participants. Groups of *Castellers*, often from other parts of Catalunya, vie with other such groups. The goal of each team is to construct the highest tower with the most spectacular gymnastic difficulty. The latter depends upon the number of individuals on each layer, as well as the type of support provided from below. Each variant form provides a difference in scoring. While there has been a drop off in participation by the young in the Sardana, the *Castellers* remain very popular throughout the various Catalan-speaking areas and each year the number of participants increases.

4. National Holidays

There are three national holidays in Catalunya that are of major importance in expressing identity with Catalan culture. These are the *Diada* on September 11, the day of *Sant Jordi* (Saint George), the patron saint of Catalunya, and the summer solstice festival of Sant Joan (Saint John). The *Diada* celebrates the defeat in 1714 of the Catalans in Barcelona by the combined forces of Castilians and the French. This holiday, rather than dwelling on defeat in order to muster hostile feelings against the combined enemies (such as the Serbian holiday that marks their defeat in Kosovo by the Turks which led to Turkish domination), focuses on this date to symbolize a new beginning for the reconstruction of Catalan culture. St. Jordi is known through the Catalan-speaking territories as the day of the rose and the book. (It will be discussed in greater length a little later.) As for the festival of Sant Joan, since 1955 it has signified the Catalan desire for recognition as a separate democratic entity as well as exemplifying freedom of expression.

5. Choral Singing

This is an important part of Catalan identity. Even small villages, such as Portbou, have their own choral groups. Apart from representing widespread participation in a shared musical tradition, chorals also serve to foster community pride. The most famous of these choral groups is the *Orfeó* of Barcelona, based in the *Palau de la Música* in the center of the old town, the *Barri Gòtic*. It is important to note that the *Palau de la Música* was designed by Domènech i Montaner, a Catalan Art Nouveau architect, in a style known as "modernisme" in Catalunya. The Palau was decorated by Lluís Bru and Eusebi Aranu. It is one of the most successful architectural monuments in Barcelona, perhaps in the world, and has been classified by UNESCO as a part of the planet's great public works of art. Also worth noting is the fact that the *Palau* was constructed as a low-priced alternative to the already existing Liceu, the elitist and expensive opera house on the Rambla de Catalunya that was built to cater to the rich of the city.

6. Barcelona Football

The Barcelona football (soccer) club is another important part of Catalan culture. It is owned by a sizeable group of large and small investors. Its president is elected by shareholders that makes the team a rather democratic institution. It is also wildly popular among fans in Catalunya who hope for championships every year, and particularly victories in games with the Real Madrid team. In 2003 a new coach, Joan Laporta, took office. His first act was to demand that all team members learn Catalan.

7. Les Caixes (Savings Banks of Catalunya)

The Savings Banks of Catalunya (*les Caixes*) are very active in the support of culture in the *Principat*. They offer free high-quality art shows as well as free concerts.

8. The Nova Cancó (New Song)

Toward the end of the Franco period, a group of young musicians and singers began to compose and sing thinly veiled songs, aimed against fascist domination. Perhaps the best known of these artists are Maria del Mar Bonet, Joan Manuel Serrat, Lluís Llach, and Raimon. All four of them continue to perform at the present time, but while Serrat still sings in Catalan, many of his current recordings are exclusively in Castilian, a decision that has increased his popularity in Latin America. Bonet has not stopped singing in Catalan but now specializes in music from the entire Mediterranean region. Raimon continues to sing both his old protest songs and musical compositions set to the poetry of two great Catalan poets—the Valencian, Ausias March, who composed his work in the fifteenth century and Salvador Espriu who lived from 1913 to 1985. Lluís Llach continues to perform militant songs that stress left-wing principles, and Catalan independence.

9. Excursionism

A feature of Catalan nationalism involves the active participation of people from all walks of life in the exploration and appreciation of the rural geography of the entire region of the *Principat*. This remains an important activity to this day. In the Catalan context, however, excursionism should not be confused with the "Naturevolk" ideal of Nazi Germany that was designed to inculcate the young with a kind of pagan alternative to other religions as well as the principles of Nazism. The focus of excursionism in Catalunya is to gain familiarity with the varied Catalan territory and to make participants aware of the ecological value of their land. In no way is this custom seen as an aggressive form of nationalism.

10. A Key Symbol

Early in this new century the followers of the right-wing nationalist party of Spain, the PP, adopted the symbol of the fighting bull as their symbol. Often one now sees the bull added to the Spanish national flag to consecrate the identification of the party with a united Spain. In 2004, a group of youths in Catalunya began to display their own symbol of Catalan identity in the form of the Catalan donkey, or "*Ruc*," actually a large and very strong breed created by the family of President Woodrow Wilson and sold in large quantities to Catalunya at the beginning of the twentieth century. For Catalans this new symbol represents their stubborn but nonviolent resistance to domination.

11. Computer Games

Most recently a movement to create computer games in Catalan has begun with the explicit goal of encouraging youngsters to use the language.

12. CAT

In the fall of 2005 the international association responsible for standardizing international symbolism on the Internet approved the category CAT to be used in Internet addresses dealing with, or written in, the Catalan language.

Conclusion

I hope the reader will agree that the historical development of Catalan nationalism fits like a hand in the glove with Benedict Anderson's analysis. The nationalist movement in Catalunya clearly began, as Anderson would claim, at the earliest in the second half of the nineteenth century, as a manifestation of the Catalan "Renaissance" with the revival of the *Jocs Florals*, a clearly cultural phenomenon that had political implications. It was not long afterwards that a new literary tradition with a range of publications in a rather standardized form of Catalan led to the recognition of a strong cultural identity. This tradition continues today. A large number of Catalan publishers produce a wide variety of books in the language including prose fiction as well as poetry, classics, and modern works, in addition to a vast array of translations from a wide range of languages. In the late nineteenth century, in addition to books, a range of Catalan magazines appeared as well, some of which had a political slant, somewhat tempered, however, by a humorous style.

As Benedict Anderson and others claim, Catalan nationalism can also be characterized as a response to industrialization that began in earnest in the same period. On the other hand, it was only when the nineteenth century was practically over that a strong political force came forward in 1892 with the creation of the *Unió Catalanista*. For the first time a range of diverse groups were united into a truly national movement. The first Catalanist political victory came shortly afterward in 1901, when the *Unió Regionalista* fused with the *Centre National Català*. This new force won an overwhelming victory in regional elections in the same year. All the candidates for the presidency of the four *diputacions* (regions into which Catalunya was divided by the central government in Madrid so that it could better control the territory) were elected. It was not until 1914 that the *Mancomunitat* was formed to bypass the separation of Catalunya into separate political entities. This unity was not long lasting, however. It was abolished in 1925 by the Spanish dictator, Primo de Rivera. The maintenance of these separate regions of the *Principat* continues, as of this writing, to be a major contention between the Catalan government and Madrid. The autonomous government of Catalunya is now preparing a new territorial division, the *"vegueries,"* to replace the four diputacions of Llieda, Girona, Barcelona, and Tarragona, with seven more logical political-geographical divisions.

Finally, it is important to remember that, as Anderson points out, and many other authors refuse to admit, nationalism and racism are not part of the same process, although they can occur together in the same context. Again, Catalunya presents a clear case in which racism plays no role in the national aspirations of the Catalan people. I do not mean to say here that there are no racists among Catalan nationalists—I have met some—but Catalanism as a movement has generally emphasized the assimilation of immigrants into Catalan culture.

Chapter Three

The People of Cerbère Speak

> Cerbère is a theatre set because someone has decided that the frontier is really a *frontier*. Between Catalunya and *el Rosselló* [French Catalunya] there is no real frontier so the French have created a false scenic one. The urbanism of the village has been consciously created to signal the difference. Because of the absent border they have created houses with roofs that are un-Catalan and markedly French. If you go further north into France (Banyuls, Collioure) you will find typical Catalan roofs like the roofs found throughout the south. But in Cerbère they have to be different.
>
> —Josep Pla, "How the French Established Their Identity in Cerbère." *Contraband* i Altres Narracions. Barcelona: Edicions la Butxaca: III.

In 1996 we taped interviews with thirty-six people in Cerbère. The sample used was neither random nor stratified. Most interviews were conducted with single individuals, but there are a few exceptions to this general rule. We taped some married couples together, and on three occasions, we interviewed groups of four individuals with more or less the same set of interests or social standing. I have selected nineteen of these interviews for inclusion in this chapter. Like the Portbou interviews these have been edited for continuity; furthermore, certain material, either too sensitive to include or irrelevant for this study, has been eliminated. What I have chosen to include from the tapes illustrates the general malaise felt in Cerbère concerning the well-being of the village in the future given the changes that have occurred in recent years. The interviews also provide some insight into the wide range of viewpoints in the village concerning personal identity in the context of possible Spanish, French, and Catalan identity. In Cerbère, somewhat unlike Portbou, we favored interviews with people who were associated with its stable population and who more or less saw themselves connected to Catalan culture. I do, however, include some interviews that reflect the feelings of the transient population of the village as well.

* * *

Interviewee 1. Man, retired, born in Cerbère, approximately seventy-five years at time of interview.

I was born in Cerbère. We spoke Catalan at home but were not allowed to speak it in school. In the playground we spoke Catalan. My wife does not speak it nor does my son. He speaks Castilian, learned while working in Portbou where he was a railroad employee. In the past there were about six hundred workers for the railroad in Cerbère. That was before the Spanish Civil War—1934–1936. Women workers transferred fruit from Spanish to French cars. After the war there were about sixty transfer companies. Their businesses began to fall off after the Second World War. The major transfer work is now mechanized and done by a Spanish company, Transfesa. During the Second World War there was still commerce from Spain. Much of it was Spanish coal; there was also oil and copra. [Did many refugees come to Cerbère after the Civil War?] There were many who stayed here quite a long time and, then, when the law in Spain changed they were allowed to go home. Many of them did. Most of the refugees here were Catalans, many from Portbou. Some of them became French. In 1976 I retired from the railroad at the age of fifty-five. I worked in the office. [Today, in the actual population how many are Catalan more or less?] It has changed a great deal. There are many people who have come from the north of France. Cerbère is a French village. During the war, when the school in Portbou was bombarded by the fascists, the children came here for school. Cerbère was also bombarded. In 1878, the railroad station was opened. [What happened here to mark the centennial in 1978?] To celebrate the event we prepared a feast for the chiefs of the SNCF [the French National Railroad] but when no one showed up everyone went on to Portbou where they had prepared their own celebration. [Was it important for the people of Cerbère to celebrate the building of the station?] Yes, the railroad is what made the village. The real natives of Cerbère are very few in number, about two hundred fifty. The *pieds noirs* [French Algerians who were forced to leave Algeria in the 1960's when that country became independent] began to arrive here in 1962–1964 in large numbers and most stayed, primarily to work for the railroad. Many spoke Spanish, but not Catalan. Only the old speak Catalan here. I was deported during the war to Germany and learned their language. I was assigned to work for the railroad, but as an ordinary worker on the trains rather than in an office. [Why did the Hotel Belvedere close?] There was not enough tourism here to support it. [In the old days who were the people who spoke Catalan?] Most of the people born here spoke it. [Is there a difference between the Catalan of Portbou and here in Cerbère?] Yes, there are quite a few vocabulary differences. [Do you see the people of Portbou as Spanish or Catalan?] Catalan. Catalunya is big and very rich; there is Barcelona. [Do you have friends in Portbou?] Yes, the priest and many others. [He tells us about the bulldozer that was lent to Portbou to improve the road between the villages and how the French government made a fuss over it.] [When does the tourist season begin here?] It starts in July. [Are there associations here that villagers can participate in?] There is a retired persons club. There is the nautical club also but that's a private business. There is very little work here. [How many food stores were there here before the crisis?] Seven or eight. There was even a print shop here until about only fifty or sixty years ago. [What were work conditions like?] I had my first vacation under the government of the United Front in 1936. The United Front passed many social laws. At first we had twenty-one days of vacation. I began work at the age of thirteen and never had a vacation before. When I was a child my brother and I collected the oranges that fell out of the baskets of the workers as they carried them to and from the trains. [What new businesses have been created here recently?] The clinic [for severely handicapped people] created about two hundred fifty jobs, but very few people from Cerbère have the qualifications to work there. There are people who live in Cerbère who work in Banyuls and Port Vendres. [Is there adequate housing in the village?] The houses that belonged to the railroad were bought by a private company and are now for sale. The large apartment house near the station still belongs to the railroad company. [What can these sales do for Cerbère?] Not much, but they will pay taxes to the villages. [What other businesses exist here?] There is a Friday market every week when traveling merchants sell food and clothing. [What about the restaurants and hotels?] In the winter the hotels and most of the restaurants are closed. Summer is the only busy time.

Interviewees 2 and 3. Men, the first a member of the village council who retired from work three years ago; the other works for the tourist agency.

Interviewee 2. [Where are you from?] My parents and my grandparents are French Catalans. I was born in the Ariège [a non-Catalan province in France]. [Do you speak Catalan?] I speak Catalan and have spoken it since my childhood. When I was little in our village in the

Cerdanya [a Catalan-speaking part of French Catalunya] almost everyone spoke Catalan. Then many people left and Catalan has been partially lost. Here I speak it with the older people who do speak it. Many Cerbère natives have left and now we have many *pieds noirs* here. Now in Collioure there is a [Catalan private] school and in Perpignan there is Catalan in some of the public schools, the Arrels [in which not all courses are taught in the language]. On the initiative of Porbou, people from here go there to learn Catalan. In reality the course teaches only basic Catalan, enough to ask where the market is, etc. There are also people in Portbou who don't speak Catalan because they are from other regions of Spain. [What nationality do you consider yourself?] I am French. Catalunya is France for me. I don't think that independence is possible. In Portbou they have another opinion. Many of them want Catalunya to be independent. It is important to keep the culture alive here. If I want to buy something I go where I want to. [Is it really cheaper there?] Hotels and restaurants are cheaper there. [How is the economy doing here?] There are less jobs since Transfesa came and the merchandise began to be transferred by machine. Before the work was more labor intensive with hand transport from one train to another. The mechanized transfer system began before 1975. With this system six to eight people can now do the work in a few minutes that would have taken all day with many workers. Work was done in spite of the weather. There were also fourteen to fifteen commercial houses that hired women to transfer merchandise. They too began to disappear with the arrival of Transfesa. When my father was here—he retired in 1961—there were sixty customs agents. [How did the inspection system work?] The trains were sealed with lead until the customs agent came to investigate the contents. The loss of all these businesses as well as the customs agents and rail workers meant that entire families left the village. The loss of population had a great influence on commerce. Before there were eight or nine grocery stores, now there are only one and a half and one of these will close in a few months. Ten years ago there were about fifteen or sixteen customs agencies. When families leave, the school's population is reduced in size also and some teachers have to be transferred. In the apartment block for rail workers there were sixty rented apartments. Now it's half empty. If the young don't leave there is little for them to do in the village. One of my sons works for the SNCF, but in Paris, and my daughter's husband works in the north of France. People don't necessarily want to leave because there is a good climate here. [Do you have any social functions in the village?] Yes. I'm responsible for the festival committee and am president of the sports association of the rail workers, which now includes nonrail workers. There is basketball and football in the winter. There's skiing in the winter too and swimming in the [outdoor] pool in the summer. The gym is about thirty years old. There's also European bowling and tennis. The village is too small to have a team for every sport so a soccer team was organized with Banyuls, Collioure, and Port Vendres. In the summer there are balls in different places in Cerbère. [Are there relations with other villages in Catalunya?] Yes, there are boating competitions. Each section has its own president and committee. I'm a member of the nautical club and every spring we put piers in the water and take them up in the fall with the arrival of bad weather. There are about one hundred forty boats that dock here in the summer. [What do you think will happen when the new fast train bypasses Cerbère?] I don't think that the two national governments (France and Spain) who pay for the TGV [the high speed train] will want freight traffic to pass elsewhere. But if that happens Cerbère may also lose its international transport. We may lose Cerbère-Geneva, Cerbère-Italy, and Cerbère-Paris. If there are too few passengers then the number of people who work to repair the cars will also diminish in number. In winter there are very few people here. [Can tourism save the situation?] Cerbère cannot live exclusively on tourism. There is very little hotel space and the Dorade [the one medium-sized hotel in the village] is closed all winter. The Vigi [another hotel with very few rooms] also closes in the winter as well as the other small hotel. [What will become of the businesses?] There are two bakers who will probably continue in business. [Do you remember the Belvedere in its heyday?] Yes, it had a very lovely cinema. The hotel closed in the sixties. [What else has hurt the village?] TV killed many of the friendly ties that held people together. The mentality here in Cerbère has changed. Before people came here and died here. Now people come for a few years to work and then go home. In the two or three years they stay they don't adapt to the culture. In the past many who worked for the railroad stayed longer and became used to village life.

Interviewee 3. I am of Spanish origin. I work with someone from Portbou. My grandparents came here in the middle of the nineteenth century. I speak Spanish. I take care of the tourist office. [What do you consider the most important aspect of Cerbère?] The identity of Cerbère was the railroad. We now have to focus on tourism. The attraction here is the fact the area is still not overbuilt. We have to convince people that it's important to have that kind of tourism. If we do that the season can be extended from Easter to November. One of the things that can be exploited is the underwater park [a nature preserve] that is magnificent. This is a very lovely coast and it already has a certain amount of government protection. It's a good thing that the current generation is interested in protecting the environment. [Do you know of some young couples who have decided to live in Cerbère?] That's the big problem. We don't have enough good housing. [What about retired people?] Yes, we can try to get them. We have many residents who consider this to be *their* village, but we also need new people. [What about combining your forces with Portbou and other Catalan villages in the other side of the border?] We're trying to do that in planning for tourism. If we can organize politically, cooperation in the area of tourism will result and we will be better listened to by Europe. [Do you think the two national governments want to cooperate?] The French government has some problems. In principle we are regulated by European laws and negotiate on that basis, but the French government is less willing to accept Europe than the Spanish government. [What language do you use?] I speak Spanish well and most of the people in the government in Portbou speak French. We do not have problems communicating with one another. I am hopeful. I have a project on which I have been working for eight years. [What is your project?] It is tied to tourism and to have visitors here at least six months of the year. It's related to the Mediterranean plan, in particular a well-organized program of activities taking advantage of the natural area. Finances are the least of the problems. First we have to have a viable program. For finances each person will have to provide his or her own. What we hope to do is to provide a stimulus.

Interviewee 4. Man, retired, eighty-nine years at time of interview! Worked as a partner in a *transitor* agency. He died a few months after this interview.

[Can you give us your family history?] I was born on the twenty-second of November 1907, the same day as General de Gaulle. I was "fabricated" in Cerbère where my family lived at the time. But I was born in Llançà because my mother wanted us to be born there. My father, who had come here as a fisherman, did not allow us to follow in his footsteps. We always spoke Catalan at home. [In Cerbère he still speaks Catalan with friends of the same age group.] I was schooled in France. In nineteen twenty-one I took a year of Spanish in school in Llançà when the language of the school was Castilian. None of the teachers spoke Catalan. When I came back I was employed by my brother, a *transitor* (handled paper work dealing with transnational commerce). He began his enterprise in 1910. He had a partner and they worked in both Portbou and Cerbère in the same transfer company. Calcina and Bosch was the name of the company. I worked as the negotiator between the seller and the buyer. At the time there were thirty-eight *transitor* companies. That was between the twenties and the late thirties. I know by heart the names of all those who worked for the railroad, the customs, the women *transboarders* (who, exclusively, transferred the merchandise between trains by hand), and the *transitors*, two thousand and seven persons. Before the railroad, Cerbère and all the other people living on the coast made their living either from fishing or farming. There was also smuggling. The area was very poor. In 1879 everything changed with the railroad.

I knew Mr. ——, one of the legendary early entrepreneurs of the village. Mr. —— earned a lot of money at the *Buffet de la Gare* [the café at the train station]. He invented the food basket for travelers. In it he put a little bread, a bit of cheese, some chicken, and a bottle of "mineral" water. He also ran the money change office in the town. In the evening, after work, he boarded the train to Spain and changed money for the passengers between stops. While working on an important building in the village he had the bad luck to fall off a ladder and was incapacitated.

[Did people like him?] Yes, he was well liked. In 1945 the Germans gave an order to burn down the station just before they left, and he saved it. [Where did you live at the time?] Up to 1948 I lived near the tunnel [for the train tracks between the French and Spanish borders]. The Germans were here for about two years. Some of the rail workers joined the resistance against them. In particular they served as *passeurs* [aiding refugees to cross the border into Spain]. During the war the women were evacuated from the village. It was the French collaborators who made up the list for the Germans. [Why were they evacuated?] The Germans did not want to deal with them.

[How did business go for you after the Spanish Civil War?] We had to close down for a long time. My brother was put on the "red" list by Franco and, as a result, we could not work even here since we had to deal with the Spanish government. [What did you do during that period?] A friend of my brother invited us to work in Narbonne [in southern France outside of French Catalunya]. [Why did Franco put you on the red list?] Because we were Republicans, and they knew it. Later we began to work in the profession again. [How many workers did you have?] Six employees. In Cerbère there were in all thirty-eight *transitors*. Each one directed a team of five women. The latter were hired by the day. There were trains full of potatoes, later fruit of all kinds—oranges in the winter. The potato baskets weighed fifty kilos. There were also artichokes.

[Did the women look for this work?] Yes, some of them earned more than their husbands who were customs agents or rail workers. For Christmas the women often worked until midnight. The station was an "ant colony." Once there were three hundred freight cars filled with oranges. Nineteen tons! These were emptied and loaded by two hundred women in one day. Each worker was given a coupon for each basket unloaded. During that time Cerbère was very rich. There was no unemployment. As soon as the children got their school certificate they had a job. There were four choices for jobs: customs, *transitors*, rail workers, and *transboarders*. [When did things begin to change?] Ruin came with the suppression of hand transfers between trains. Now it's all automatic and is done by Transfesa, an international company run from Madrid and Paris. Transfesa began in the thirties [but, we say, the monument in the town lists the last transfers occurred in 1966]. Yes, but very few trains were *transboarded* toward the end, only those that were unequipped to make the change by machine to the other rail system. Transfesa was the death of Cerbère. The women lost work and the *transitors* had less work as well because the railroad changed the system. They did make some money under the table. During the good times women came from as far away as Argelès [a village not far from Banyuls] to work as transboarders, and on the Spanish side just as far away as Vilajuiga [a small village three train stops from Portbou] The workers went home by train after their work. There were very good relations between Portbou and Cerbère. During good times some of the merchandise had to be transboarded at Hendaye [on the Atlantic coast near the border between France and Spain] because there was too much traffic for here. During the Spanish Civil War all the traffic stopped. The village council here took out a loan to help the unemployed. I went to Spain for nine months to fight for the Republic [he participated in the terrible Battle of the Ebre (Ebro in Spanish)]. [You came back to Cerbère after the war?] Yes, but the war was very hard. In the headlines of a Spanish paper we were called "the Gods of the Ebre." There were many victims in that battle. It was atrocious. In one night on a hill that we occupied we counted the impact of sixteen thousand shells that fell on our position. The next day our company, originally made up of one hundred twenty men, had only seven left alive! Among these seven there was a sergeant and a corporal. On this mountain! [Did you go to war with comrades from Cerbère or alone?] I left alone. My brother had already gone over. Three or four men from here were killed in the war. We came back the last day, the twelfth of February, in the evening. Here is a photo of all those trucks. I was asked by an officer at the border if I was French and was told that we were stopped because we had to give up our arms. At Le Pertus [the present border for the new autoroute between Spain and France], because Franco's troupes were just behind, the French gave the order to let the Republicans pass without a control. [At the time you felt Spanish?] Yes, and Republican. When we passed the French office at the frontier we could not hold back our tears. The commandant of my brigade was there. My mother died, forty-eight hours later, just forty-eight hours after our arrival. We went to Narbonne to work, and since I had studied agriculture a bit I started a garden in order to feed the family. I planted tomatoes and potatoes. The people in Narbonne spoke Occitan [the regional language still used at the time]. Those who spoke Catalan spoke a different dialect from the one here. [What did you do for leisure before the war?] I enjoyed sports. The one I liked best was swimming. In 1925, I saved a young man from drowning. He became the deputy mayor and for nine months he was minister of industry in the French government. [He then tells us that his grandfather died in 1913.] [What is your nationality?] I am French. I am Catalan and French.

Interviewee 5. Man, retired, was a fisherman in summer and in winter worked in the vineyards. After the Second World War he joined the railroad.

[Where were you born?] I came to Cerbère when I was six months old. I was born on the Spanish side. One parent was Spanish and the other French. [Why did they come here?] To find work. There were many jobs here at the time. My father was a grape farmer. [Have you lived here all your life?] Yes, there were many fishermen when I began to work. We all fished for sardines. [What did you do during the Spanish Civil War?] I fought on the Republican side. [Tell us something about Cerbère's economy in the old days.] It was the station that made Cerbère. Before the station there were only fishermen here and only in summer. In winter they left for Banyuls. [Do you speak Catalan?] Yes I speak Catalan, and a bit of Spanish as well as French. [Was Catalan spoken here much in the old days?] Yes, but there were many functionaries who did not speak it, though many of them did learn the language after being here a while. Certain young people spoke Catalan during recess in school, but it was against the rules to do so. When I was in Barcelona during Franco, Catalan was also forbidden and one had to pay attention with whom one spoke it. I was in a hotel and spoke to a man in Catalan who turned out to be a police inspector. He arrested me immediately. It was a very serious offence to speak Catalan in Spain during Franco. Catalan was not forbidden here, except in school, but there was pressure against using it. [Now when you speak with your friends what language do you use?] That depends—but when it's Catalan friends we speak in Catalan. Here there are many of the older people who speak it. In the large cities all around France there is an association of Catalans, the *Amics de*

Catalunya. There is one in Perpignan, but since Cerbère *is* in Catalunya we really don't need one. [Is it only the language or is it more than the language that makes one a Catalan?] It's mostly the language—we are not like the Basques [many Basques no longer speak their language but are still adamantly nationalist]. [What about food?] No, Catalan food is similar to food in France. The specialty is paella. [What about Catalan festivals?] There are some in summer. There are dancers of the Sardana who come here, and some other Catalan dances are danced as well. Before, everyone participated, but now it's mostly for tourists. [Are you happier now than in the old times?] We live better now but the mentality has changed. Now there are all kinds of social problems. It's like life in all the big cities. In the past it was more familial. [Were there differences between the rich and the poor here?] Yes, the rich were the *transitors*. They earned a lot of money. [He then speaks about the transboarders and why they were all women.] The women did this work because a lot of the traffic was loose fruit. They worked with baskets. The men did harder work, although it was not easy for the women who were paid by the ton. The heavier the day's work, the more they were paid. But they were not well paid. Considering the hard work they did they were exploited. Sometimes they were obliged to work twelve to seventeen hours at a time. [Were there unions?] The rail workers were in part unionized. But the railroad was private and there were four different companies. The unions were very weak before the war. At first they were created by Petain in the forties when the Free Zone existed. [Is Cerbère a village on the left or on the right?] It's traditionally socialist. [Did the unions get bigger when the system was nationalized?] Yes, and there were four unions. There was also a syndicate of the cadres. The women were never syndicated but they did go on strike. They were supported by the men and they usually were granted some of their demands. [Are there women alive here who were transboarders?] There are some, but I don't know who they are. My mother was one. [What about schooling?] School was obligatory up to sixteen years of age. [Did some go on to high school?] Yes, but I did not want to go. Some did and became professionals. I preferred to fish than go to school, and our parents did not push us to get educated. School at the time was not free and there were no scholarships. Before there were no governmental family allowances like there are now. [When the Germans arrived here what was the atmosphere?] They came here on the sixteenth of November 1942. Many of the young men at the time went to Spain. The Germans are said to have behaved badly but I did not have much experience with them. After a while Cerbère was full of

German rail workers, German customs officials, etc. I left for a few months in December of the same year. A group of young men from here left for Spain. Many crossed the frontier without problems because most of the customs officials were okay. Some, however, were taken and shot. There were people here who helped them escape to the other side. [Were there Jews and others who also escaped?] Yes, this was one of the passage places. When it was the Free Zone many Jews came here and when the Germans came, many of them escaped to Spain. There are some who are now in Israel and others who now live in Paris. Many of them have kept up contact with the people here.

Interviewee 6. Man, retired, in his sixties. Has a vineyard and makes his own wine. When we interviewed him we noted that his casks for the grapes were painted in Catalan colors.

[Where were you born?] I was born in Cerbère, but my mother was from Llancà. When did you learn to speak Catalan? I never had to *learn* Catalan. I had to learn French. I have always considered myself Catalan. My father was one of the founders of the Barcelona football team [affectionately known as Barça]. I have the right to enter the Barcelona stadium area just the way the players can. [What does it mean to be Catalan in France?] The country is Catalan just up to Narbonne and in Spain just to the line with València. For us there has never been a frontier. Even during the wars. We always had a great deal of contact with both sides of the frontier. Have you seen the monument to the dead of the First World War here in Cerbère [he asks]. It is for the children of both Cerbère and Portbou. [Did many from Cerbère fight in the Spanish Civil War?] There were many who fought against Franco. I have a cousin, for example, who joined the Republicans. [Are there many people who feel the way you do about being a Catalan?] Not many. When I was in school the teachers hit me on my fingers when they caught me speaking Catalan. Now at least one can speak Catalan here. I am a *Sardanist* [dancer of the Sardana] at the age of seventy. But when the Sardana music was played in Cerbère twenty years ago there were those who made fun of it. Now the same people dance it. I did not change but they did. There is a bit of folklore in it, but I think also that the people, now, are proud to have a capital in Barcelona. This has helped the spirit of cooperation between Cerbère and Portbou, but our current mayor feels less Catalan than the last one. In the old days lots of merchandise was transferred here, and there was lots of work. People came here from both sides of the border to work for my father. Most were from Llancà, Colera, and Portbou. My father had a business that provided work for about sixty transboarders. He was the middleman between the *transitors* and the transboarders. Of the five women who worked as transboarders, four filled the baskets and one carried it across to the other train. My wife was one of the women who worked on the bridge between the trains. After a while my father changed his business into a cooperative. The men and women workers ran the coop. [Was there a resistance against the coop?] No, it was well accepted. After the war my father recreated the business with his sons. But by then Transfesa had been created and work had changed. Cerbère has changed very much with the mechanization of the transfer of the trains. [Do you know much about the strikes by the women?] I remember two strikes of transboarders. My father remembered at least fourteen of them. The strikes were not long except the big one that is described in the thesis. [The study of the two villages mentioned in chapter one. He notes his interest and that of other residents of Cerbère.] There are seven copies in Cerbère. Some of us even went to Aix en Provence for the defense. The mayor went and I accompanied him as adjunct mayor.

Interviewee 7. Woman, early fifties, teacher in the village nursery school.

[Where were you born?] I was born near Perpignan. My husband was born in Cerbère. He is Catalan, and speaks Catalan and Spanish as well as French. He visits Portbou often. His mother was from the south of Spain. He loves to go to Portbou to play a Catalan card game. He works for the Post Office. My family spoke Catalan at home. It's my maternal language. The dialect here is a bit different than in the *Principat*. When I went to school, classes were, of course, in French. In school I never saw how Catalan was written. I was happy to discover that I could read it in spite of sound differences. I don't write the language. With our son we speak French but he took the Catalan option in high school. He also speaks Catalan with other members of the family. He learned it easily. He is now twenty-eight years old. He works in Banyuls. The school is losing students with the drop in population. When I began teaching in the sixties there were three classes here. Now we have only two. And some of the children are from elsewhere. I have some younger children from Banyuls in my class who are two years of age since that village [school] takes kids only from the age of three. The age to begin elementary school is six in France. [How many are there in the two classes?] I have twenty-seven in my class and my colleague has twenty-two in hers. [Do you expose the children to the Catalan language at all?] Yes, I teach them to sing Catalan songs [she gives us an example]. I take a group of six to eight children and we do little scenes in Catalan twice a week, one group on Mondays and the other on Thursdays. We have a little Moroccan in the class, and even she participates. [Why do you do that?] I do it because I love the Catalan language and it's my roots. One of the songs we sing in class was taught to me by my grandfather when I was a child. When I went to school we were not allowed speak Catalan. The older generation was forbidden to speak it even on the playground. I teach Catalan also because it works well as the first contact the children have with another language. It's difficult since there are sounds in Catalan that don't exist in French. Each year there is a contest in the nursery school. We chose Santa Claus as the subject one year. I translated a story from French into Catalan and then I taught it to the students and we put it on a tape. We followed the rules for the competition and we won the first prize from the *Generalitat* of Catalunya. [This is an example of how the Catalan government in Barcelona aids the propagation of the Catalan languages in other regions.] There were quite a few schools that participated. [Would you please read some of it in Catalan to us?] [She then reads some of the text that is on the tape.] [Do you participate in Carnival with the students?] Yes, we do, and we celebrate it in the Catalan style and the songs are sung in Catalan. We perform in the main village square. [What are the Catalan traditions that exist here?] There is the Sardana. [Is there Catalan cooking?] Yes I like to cook that way also. [Do you speak Catalan with many people in Cerbère?] I speak with most of the people here in French, but with some older folks I speak in Catalan. [Is Cerbère much different than it was when you came to the village?] Yes, there were many more people. I had the feeling of being part of a large family. It had, and still has, a very congenial atmosphere. There are less stores now than before. We generally go to Portbou to visit the cafés. Here the cafés close in winter. Before they didn't. [Do you shop in Portbou?] Partly there and partly in Banyuls. [Do you think there could be better relations between the two villages?] I don't know. My niece was in the exchange program. Her mother is from Portbou and the entire family speaks three languages: Spanish, Catalan, and French. Their child is ten years of age. She also speaks three languages. The child's grandparents on the mother's side live in Portbou. One day we were in the car with her on the way to Portbou. Up to the frontier the child spoke in French and from the border on she spoke in Catalan. Now she speaks Catalan. For a while she mixed the three languages. "*En casa mia j'ai une poupée.* [I have a doll in my house.]" [When you retire where will you go?] I will stay here. I feel like a native of Cerbère. We have a house under the gas station facing the sea. My son lives there now. He runs a snack bar on the same property that is only open in the summer. [Is there an active social life here in Cerbère?] There are sports: swimming, basketball, tennis, and French bowling too. The older men fish in the better weather. My husband belongs to the bowling clubs in both Portbou and Cerbère. The club in Portbou is much more active than the one here. I have always liked Portbou because it's more active than here. There are always people in the cafés on the beachfront and in good weather throughout the year the families take walks on the seaside and the rambla. We really like to go to Portbou. There are more people there who originated in the village whereas here there are more transients. Portbou people are warmer than the people here. In Portbou the people are out late at night. They eat at ten, eleven, even twelve o'clock. [Do have much contact with the families of the railworkers?] Yes, many are the parents of my students. [Do you belong to an association?] No. I give a lot of time to the school. I like to read and sew. I have helped Mrs. Carcel [the woman who sews for the villagers] prepare costumes for the festival. [Are there many ties between Portbou and Cerbère?] No, I don't think so. [Do you have the impression that the village is Catalan or French or both?] I ask myself: Am I Catalan? Am I French? Am I European? I find it difficult to respond. [We tell her that we find this answer interesting, and say to her: If you were in Paris you would have no problem responding that you are French.] [What about your parents? What would they have answered to this question?] Perhaps that they were French and Catalan. My husband, when he retires, would like to live in Portbou. But he is also very attached to this village. His mother came to Cerbère during the Civil War and he came here with her. Her family was quite well off. And they had to flee here. They left all their things in Spain in the south. My son wanted to take his vacation in Spain to find the village of his grandmother. He found the place where her house had been, but now there is a parking lot. So he never went back.

Interviewee 8. Woman, in her late sixties, hairdresser.

[Where were you born?] I was born in Cerbère in 1928 and have never left it. [Where were your parents from?] My mother was also born in the village and her father was from Northern Catalunya (Roussillon) near Perpignan. He worked for a *transitor* and then he worked for the railroad. My parents were Catalan. [Did they speak the language?] They both spoke Catalan at home, but in school it was not allowed. In the house they spoke Catalan with one another and with us in French. Now people are trying to revive the language. [What do you think of that effort?] I think it's magnificent. My brother says he is more Catalan than French. I agree with my brother. Catalan culture is so beautiful. [What do you mean by Catalan culture?] Many things. We have ideas that are a little different, a bit like the Basque. We are from

Northern Catalunya. The south [the *Principat*] is not the same as here. They have another way of looking at life. [How so?] They are perhaps more "*attachant*" but they are too proud. The Catalan here is very different from that in Portbou—there are many words that are different. Even the Catalan in Banyuls is different from here. I learned Catalan as a schoolchild even though it was forbidden. They didn't punish us but they had strict rules against our speaking it. [What was Cerbère like in the old days?] It was the richest town on the coast. People came from Perpignan to work here. There were many more people. Aside from the workers related to the railroad commerce there were three hairdressers, five butchers, and many grocery stores. That lasted until Transfesa began. Cerbère continued to be prosperous until 1968. My mother was a transboarder. She emptied the baskets. One cannot imagine the number of tons she transported. She earned a bit more than the other women. There were five or six tons in each wagon. When she came home she was in a state! Around Christmas she often worked up to midnight. It was slavery. They were not well paid. That's how the *transitors* made their money—on the backs of the women. I even had an uncle who was a *transitor*. At the time it was normal. We had to have them to do all the paperwork. [Will you tell us a little about your childhood?] We were three children in the house. Mother had to come back and feed us. On Sunday the women didn't work. She would then go to the public washbasin to wash the clothes. She was sixty when she stopped carrying oranges.

[Now how do you see Cerbère?] It's a village that is dying. I don't think it's possible to revive it. There is an opposition that thinks it's possible but I don't think so. [Is it possible for the two villages to help one another?] They are in the same situation. But there were many more *transitors* in Cerbère than in Portbou. Also they have much more tourism than we do. It's cheaper there. There are more shops in Portbou. People here who retire cannot sell their shops. There are too few people here. I don't think tourism can help and there are not enough vineyards. The wine industry is much more important in Banyuls. [Is there adequate housing in Cerbère?] There is the new project, the Solane, on the hill. Many young couples have built there and that keeps some people here. It's a special situation because we have so many people who pass through and are from different regions and don't speak Catalan. Now there are few workers even for Transfesa. There used to be many more. [How do you feel about all this?] It's very painful for me. This is my village and I want to die here. I have had an active life here. I have my family here. [Do you think that tourism can help?] Cerbère is very small. Where are you going to put more than seventy-five tourists?

The old Hotel Central was bought by the town when it went out of business because it was not very comfortable. Now tourists want all kinds of comfort during their stay. Soon there will only be strangers here; it will no longer be a Catalan village. All the apartments in the buildings belonging to the railroad are being sold. Tourists are taking advantage of the low prices. I have nothing against those people, they are French after all, but these apartments are empty all winter long and that's the problem. The customs bureau is also selling its apartments. The Ministry of Finance has bought apartments near the church for vacationers. The SNCF also rents small apartments in the same neighborhood. [What are the associations that exist here?] There is the Sardana group, the gymnastics club, and there used to be a photo club. Then there is the sewing club that used to be in the City Hall and has now moved. There is a library next to the station. [Are you going to retire soon?] I should have retired eight years ago but I have continued to help my worker keep the shop going. I have no children. I was not able to have any. [Does anyone else in your family speak Catalan?] The children of my brother do not speak Catalan. It's great that Madame ——— teaches some Catalan in the school. She is perhaps the best Catalan speaker in the village. We have always spoken Catalan with her. We should have created a Catalan circle but we never had the time. Now there are few Catalans in Cerbère. [Are there others as dedicated to Catalan culture as Mr. ———? One of our interviewees.] No, there are very few. The Spanish who live in Portbou—they all speak Catalan. [As we shall see in the next chapter this is far from the case.] But few, if any, of the French here speak it. In Banyuls there are many Catalan speakers, and even the children speak it. The grape industry in Banyuls kept Catalans at home in their own village. Here I would almost feel out of place speaking Catalan. It's sad that even the doctors and the dentist here don't speak Catalan. The new mayor is now studying the language, but he is not a Catalan. The gendarmes here do not speak it. I have never seen a single gendarme here who spoke the language.

Interviewees 9. and 10. A man and his wife in their late thirties or early forties. They operate a shop in the village.

[Where were you born?] In Cerbère but my parents are Catalan, not Spanish. [His wife adds,] his father is from Porbou and his mother was born in Rivesaltes [near Perpignan. The man continues]. My father came here when he was six with his father in 1925. My grandfather was a shopkeeper. At that time there were many people here and many stores. There were no large shopping malls so people did their shopping here. Even people from Spain came to shop for quality goods. I learned Catalan from my grandfather. I spoke French with my father.

That's because he spoke French with my mother. I still speak Catalan because we have many friends in Portbou. When I was in school Catalan was offered as an option. It gave some extra points for the high school diploma. I can now read Catalan with facility. My wife never spoke Catalan, but she can understand it and Spanish as well. [Are there people here who have relations with people in Portbou?] Yes, there is a nightclub in Portbou that is frequented by people from Cerbère and some also go there to shop and to eat in the restaurants. Many intermarriages between the two villages have occurred. When this happens most of the couples come here to live. Personally, I like the ambiance in Portbou. I also enjoy the songs of Lluís Llach [a left wing Catalan singer who was banned during Franco and is still popular in Catalunya]. The people from Portbou who settle here integrate well into the life here. [How would you define being Catalan?] In the *Principat* the people are proud of being Catalan. Catalunya was rich and its people are hard workers. The same spirit doesn't exist in France. Here, in Northern Catalunya the people are not nationalists, and very few speak Catalan. I speak Catalan only with the old people in Cerbère and I get pleasure from using the language. To speak a language is important—it's a heritage. [We ask the man's wife.] [How do you see the future of Cerbère?] We have arrived as low as we can get and now we can only go upward. Many people who work here prefer to live elsewhere and commute. It's very difficult to find housing. The politicians here have neglected appropriate housing for workers. The only things available are low rent apartments and people prefer to own houses. It's also the case here that, because of the terrain, you have to construct as tall as you can on a small space. Also, once elementary school is over, kids have to go to Céret [in Northern Catalunya in the interior of the department] or Perpignan for middle and high school. Also many people who live here prefer to shop in Perpignan. We now have a new mayor who may unblock the situation. We had a mayor for thirty years who began very well. He developed the sewer system. He was very popular. He was rich and didn't have to work so he spent a great deal of time performing his duties. He was able to develop his businesses and many people now work for him. [Do you have a vineyard?] I would like to but I don't have enough time at present. [What about the Belvedere?] It will need a great deal of financing and thus help from the government. The owner is trying to do something with it. When you arrive it's the first thing you see and also it has an important history. Now it's falling apart. I don't think most people recognize the value of the building. [Do you know why the cooperative school project between Cerbère and Portbou failed?]

There were problems of scheduling classes with Spain because their vacations don't correspond to ours. They have many more religious holidays than the French. The daily routine is also different. And they eat at different times. All this makes things difficult.

Interviewees 11, 12, 13, and 14. Two women schoolteachers, a male teacher, and a man who is a former official of the Regional Educational Department.

Teacher number one, the oldest of the three: [Where is your family from?] My mother was Spanish but not Catalan. My father was from Carcassonne [a town outside of the French Catalan department] but the family language was Catalan. When I started school my parents tried to speak French with me to help out with my language learning. The teachers in the school did not allow the students to speak in Catalan. [The former official] My name is ——. I am a Catalanist. I have spoken Catalan since childhood. Both my parents were from a small town in the region. Many people here do speak Catalan, and in Banyuls, even the younger ones do. Here a few young people also speak it but only with one another. [What do you know about the cooperative school project between Cerbère and Portbou? [Teacher number one answers.] Catalan courses in Portbou were conceived of by the village council there and were also supported by the Catalan government in Barcelona. The last mayors of both villages developed the program. Spanish was to be taught one hour a week and considered a foreign language. The first year we did it only with the twelve-year-old children. There were three teachers involved. The Catalan and Spanish teachers were from Portbou and I was responsible for teaching in French in Portbou. We divided the students into three mixed groups, half from one village and half from the other. The second year of the program we added younger children and another teacher. I taught French in Portbou and —— taught it here in Cerbère. By the end of the program we had three teachers in Portbou and two here with three classes. After that, at the end of the 1990s, there were elections in Portbou and the new village council did not wish to continue the program. We had to pay for transportation between the two villages and all the scholastic material had to be provided by the teachers. That was very discouraging. In Portbou the parents get to appoint the school director. The director who had supported the project resigned and the new one who wished to continue with the project was forced by the parents to stop it. The parents in Portbou felt that English was more important than French as the second foreign language [Castilian was the first "foreign" language taught]. The people in Portbou were also not very interested in the Catalan part of the program since the real Catalans were a minority in the village. [Teacher number two introduces herself.] My name is ———. I am a Northern Catalan. The spelling of my name, which is Catalan, was changed into French. My grandparents spoke Catalan with me when I was a child, but now I speak it very little. It's more the culture that I am attached to. I can read the language. [What do you mean by culture?] Music, tradition, dance, folklore, cooking. There are some Catalans who are stubborn about their culture. People of my age always speak in French. In my family there are some who always speak to one another in Catalan and that's something of a problem for me, but I do understand them. My boyfriend, however, does not, although he is a Catalan. My family often watches Catalan television and my boyfriend does not understand the programs. [Teacher number three takes up the conversation.] My parents are Northern Catalans. Before I came here I taught in Céret. I did my first degree in Perpignan in Catalan studies. First I was a student in art history in Montpellier and then I decided to change my interest. Catalan was available and I had always wanted to learn it. When I was a child my parents spoke Catalan with one another but French with the children. I had always heard Catalan in the house. I had some problems with pronunciation when I began seriously studying the language but little by little I got over it. The program in Perpignan goes all the way to the PhD. They offer history, linguistics, economy, and literature. I also had an Erasmus fellowship in Barcelona for four months. [Do you use your Catalan here?] I use it with my parents. I also watch Catalan TV and listen to the radio and read whatever newspaper is available. The only paper available here is a satiric journal but one can also buy the monthly, *El Temps*. [Do you see a difference between the Catalan that you learned and what is spoken here and in Portbou?] There are differences mostly in vocabulary. I also learned about the different dialects of the language. When I was young I did not even know that they spoke Catalan in Barcelona. I thought it was only spoken in French Catalunya. When I was a kid I had friends who were from Barcelona, but they were Castilian speakers. It was at the university that I discovered the extent of the language. Here in the school we speak to each other in French. I teach a bit of Catalan with the first, second, and third grade classes. [Teacher number two does join him to sing Catalan songs with the students, but her Catalan is too weak to teach the language.] [Are the parents happy with this?] We don't know, but the kids seem to enjoy it very much. In Portbou the choral sings in Catalan but here our choral sings exclusively in French. [We ask about music in Cerbère and teacher number one responds.] There was a school of music here but it closed. We tried to revive it with Banyuls but they had financial problems. We do have a woman who comes to the school once a week.

[What about folk dance?] There are two groups of retired people who dance the Sardana. Originally there was only one but it broke apart because its members did not get along. [Are any of the members of either group Catalan speakers?] Some of them speak Catalan. [Do any of you go to Portbou?] [Teacher number three answers.] I often go to the nightclub in Portbou. Other boys from Cerbère often go to Portbou to find girls in the nightclub. Some of the Cerbère boys speak Catalan. [Teacher number one, the eldest of the three, adds.] In all the mixed marriages here the Catalan speaker speaks Catalan with his or her children. There are about ten intermarriages between the villages. [For you, what is Catalan culture?] [Teacher number one responds.] It defines one's roots as with the music. [Teacher number three breaks in.] I found Catalan music rather old fashioned and the Sardana boring. [Teacher number two.] When I go back to the mountains it's a Catalan bath. The Sardana is sometimes danced three times a day during festivals there. I heard the music and saw the dance in my village from a very young age. We have friends from Nice who recreate Catalan culture there. We have special foods like the *cargolada* [a snail dish] and during the festivals in Cerbère people often wear Catalan costumes, including tourists on vacation. There is less motivation to participate on the part of boys and even the girls are less interested. When the festival for children began to look a bit ridiculous in the eyes of the participants we stopped it. [Teacher number three.] I never participated in festivals. My family did participate but it never interested me. What interests me is the part of Catalan culture that is not folkloric. When I discovered Catalan culture in Barcelona I found out that one can express it in economics, literature, through the language. In some schools the children learn to dance the Sardana. [Is there something in the Sardana that unites people?] [Teacher number one replies.] In the Sardana everyone should be admitted. There should be no distinction of race or social status. In Cerbère it's different. Here people are thrown out of the group and I don't accept that. The only rule is that one should not enter the circle by separating a couple. I have a friend from Prats de Molló who was surprised that that's the way things are done here. The Sardana is the dance of friendship. Often, at the end the dancers make one grand circle as a symbol of Catalan unity. Once, when we were dancing, a group of people joined our circle (and at the end of the dance) and we were yelled at by the other, sacrosanct, group for admitting them! The cry at the end of the dance is "Visca Catalunya!" The current members of the club have taken over the dance for themselves. They don't understand the spirit behind it. I never go now and I know other people who feel uneasy joining with them. [Do you want to form another group?] That would create two groups and that's also divisive. What we do is form a group in the school with the children.

Interviewee 15. Man, in his late thirties. Works in a local hotel.

[Where were you born?] In Cerbère, my father was the founder of the hotel. He began it in 1929 with a small restaurant during the glorious days of Cerbère. It was also the beginning of tourism here and this was the only hotel except for the Belvedere. Some people loved to stay there and sleep near the tracks with the smell of the trains. The Second World War was the beginning of the end for the village. The movie house closed in the late sixties, 1966 or 1967. I remember the last film they showed there. My father is now retired. I came back here voluntarily not long after my education as an engineer. I worked in Nigeria for two years. I have the equivalent of a Masters in business, but I couldn't go on with that kind of life. I don't feel that I need to earn a lot of money. We work here in the season for five and one half months. That gives us time for ourselves. [Are there other young people like you who have come back?] Not many. There are other young people here, but I'm not typical: I have no interest in football or other sports. [His passion is opera.] My wife and I often go to London to the opera. I don't have many interests here but we do have many friends with whom we spend time. [Do you feel as if you are fleeing Cerbère the months when you leave?] No, I adore Cerbère, although I know that many people my age think that the happiness is never where they live. [Can you tell us about some of the problems that you feel are important to the future of Cerbère?] When the clinic was built there was no housing for the workers so they had to live elsewhere. This was certainly a lost opportunity. The same is true for the railroad workers, although I must say it's also difficult to build here because of the terrain. The children here are from families that have moved in to work for the railroad. Many stay only a few years. On the other hand, the people of the village own most of the shops. The dentist is retiring this year and I don't think he will be replaced. Now Cerbère is in a spiral of decline. I live from tourism and there will always be tourism. [Do you speak Catalan?] Yes I speak it because my mother is from Spain. She came here at the beginning of the Civil War to get away from it. My grandfather (maternal) was a *transitor* in Spain but when the war started he left. He was already working in France but he did live in Portbou, before the war. My father was born in Cerbère, but his mother came from the Languedoc. [Occitan-speaking region adjacent to Roussillon] [Your father speaks Catalan?] Yes, but he speaks it badly. I was raised by my grandparents. First we all lived together. Since I was little, my grandparents spoke Catalan with me and that's how I learned it. Most of the Catalans here came from the other side of the border, in Spain. The Catalan that you read is the Catalan of Spain and the Catalan spoken here is also like the Spanish side. [Were there Germans in Cerbère during the war?] Yes, they were generally officers who lived with people of the village. They, of course, imposed themselves on their landlords and lived rent-free. [Do you also read and write Catalan?] Yes, I read it but I'm losing the ability to write. I took Catalan as an option in high school. [When you meet other Catalans do you feel anything special or is it just like meeting any other stranger?] No, it's not special for me. I don't feel out of place on the Spanish side just as I don't here. [And when you are in a non-Catalan speaking country and come across a Catalan speaker?] It's a pleasure to speak Catalan —— [their daughter] does not speak Catalan, but she does learn some songs in school. When I'm with —— or the other Catalan women in the village I ask them to speak Catalan with me. It's a language that is very lively. I enjoy speaking it. When I meet up with an old person who speaks Catalan I speak it with them to be polite as well. [You are in the minority?] Yes, most here are imported families. The railroad has four hundred workers, but some don't live here. [Most are not from here. What about the school exchange between Cerbère and Portbou?] It was destined to fail. Only a half-day per week was not enough. Also, the trip over the mountain is difficult for the kids. Then, there are so many people here who are not from French Catalunya [Roussillon] and have no interest in having their children learn the language. [What about other types of cooperation between the two villages?] It's difficult because the two cultures are so different. First, there are the eating hours. The work hours in Spain are also different. There are a lot of non-Catalans in Portbou as well. I don't really believe in cooperation across the border. [Does the underwater nature reserve bring something to the village?] It does and it does not. The interdiction forbidding entry to divers is too strict. No one has the right to swim or fish there. That has helped preserve the area and that is a good thing. In Spain divers *are* allowed to enter their reserve and here it's closed. It would be wonderful if they would create a marked path open to controlled diving. Such a change would encourage many divers to come here. [Who would have to make this decision?] It's not the region that can decide. The reserve depends upon the University of Paris. [Can you point to any other problems in Cerbère?] It is a village completely sclerosed by the railroad. The current mayor and most of the deputy mayors work for it as well as seven of fifteen members of the village council. The first economic blow here was the end of the *transitors* and it was very quick.

Interviewees 16 and 17. A man and his wife, in their forties.

[Where are you from?] I was born in Algeria and my wife is from Béziers [a city in a non-Catalan department to the east of Northern Catalunya]. I was an apprentice there on the railroad and then came here to work. [What is your job?] I am a "station visitor." [What does that entail?] We verify the security of the train just before it leaves. My job is to inspect commercial loads. The wagons arrive from Spain but the trains are organized in Cerbère. [Did you live in this house from the beginning?] No, at first we lived in the railroad apartment house. [What about other railroad workers?] Most of them live elsewhere. Only five out of twelve of us who came here at the same time from Béziers decided to stay in the village. We are happy here because we have the mountains, and the sea. And what's not here can be found in the region. [Have you noticed any changes here since you arrived?] The ambiance here has changed with the loss of population. There are less people in general. There are now a total of about one hundred twenty rail workers. When the TGV begins, the passenger traffic will end here and Cerbère will be the end of the line for local traffic. Commercial traffic will continue, but I think even that will end eventually and pass by Le Pertus [on the French-Spanish border to the West]. There is nothing that we can do about this. We hope that some work will continue here. Rail workers in Cerbère may continue to take care of maintenance. On the other hand, even the local train service may end with an installation of a local bus system. [Do many people here go to Portbou on the train?] Years ago people went to Portbou on the train and now most drive. [The wife is then asked questions.] [What is your background?] My parents were Catalan, from France. [Do you speak the language?] No, I understand it but do not speak it. My grandparents and parents spoke Catalan with one another. Now they don't speak it with us. [Her husband interjects that he is not interested in learning Catalan but adds,] some of my colleagues speak Catalan. Many of them are from Portbou, but some are from Cerbère. [Where did your children go to school?] In Cerbère. [What do you know of the school exchange program with Portbou?] We heard about it, but our children were not part of it since they finished school before it began. We think it was a good program. I [He] am for tradition. Do you feel that Cerbère is Catalan or French, or both? [He] It's Catalan, here they are trying to maintain Catalan traditions. In all the regions of France people are trying to go back to their roots. [She] I learned the Sardana with the school director. [Do you feel that Cerbère is in decline?] Yes, people often speak about the decline of the village. There is much less work here. [What is winter like?] The summer is much more enjoyable. In the winter there are many less people. People stay with their family and if they want to see a film they go to Perpignan. The restaurants here are closed in winter. The activities in the summer are a good thing. But when September arrives we appreciate the return of the calm. It's fine to have tourists here for a *few* months of the year. [He] There's hunting in the fall. In the summer it's fishing and in the winter, hunting. We have a Diane [Hunting Club] here for wild boar. Before, there were no boars in the area. Then we had quail, but they disappeared because the mountain got covered with underbrush that the birds don't like, but that's the environment that boars prefer so they took the place of the quail. [The increase in underbrush is a sign of depopulation.] I have a boat for fishing, and there are still lots of fish. Because the population here is small, there is less pollution. When the few hotels here are full, people are obliged to go somewhere else. Then there is also the undersea reserve. The entire rocky coast is well preserved. [What did the mayor do for the village?] He improved the road entering the village from Banyuls, which made arriving here much easier. Before that there were traffic jams all summer long. The mayor also built the clinic. He supports the seafront law that protects the environment. I like Cerbère and I hope it will never be disfigured like the Côte d'Azur. We have to defend the environmental law. It's always the same people who return to the village because they like the environment. I hope that Cerbère will never become like Collioure. [How would you compare Cerbère and Portbou?] I believe that the people in Portbou think like us. The two villages are similar in size and have many of the same problems. They do have more stores than Cerbère, it's like the situation in Le Pertus. On the French side there are fewer commercial establishments than on the Spanish side of the border. And the prices are cheaper than in France. [Do the railroad unions function well here?] Yes. The clinic is unionized too. Cerbère is a place where people fight for their rights. When there is a strike here there is no rail traffic. We will be participating in the general strike next week. Every month we have a meeting in our work place to discuss issues. Not everyone participates. We, for example, work often at night. The others work in the day and they have few problems. Also the meetings often take place at night and those who work in the day and live elsewhere find it difficult to participate. We are for hiring young men from the village but you have to have a certain education to be qualified for the jobs. Some were undereducated and others overeducated. We fought for the young from here to get jobs, but many went to jobs elsewhere. If one works away from his house he must pay for transportation. When the railroad needs a person from the outside they pay the transfer. Our action was positive because it added twelve young people at our station, even if they don't live here.

Interviewees 18, 19, 20, and 21. A group of young mothers who drink coffee together every school day after they drop their children off at school. They tell us that Cerbère is a mixture of Spanish, French, and Catalan and that some of the Spanish in the village have learned Catalan. One of them is Catalan from Argelès [a beach side town that is not far from Cerbère]. She has family on the Spanish side near Portbou and other family members from the south of Spain. She understands Catalan well but makes mistakes in pronunciation. If someone speaks to her in Catalan she responds in Spanish since she speaks the latter better. Her husband is from the north of France. She works for the tourist bureau and the village council. [Another of these women is from Paris.] [How long have you been here?] I have been here eight years. My husband is a Catalan from France. [Does he speak Catalan with your children?] No, only French. [What does your husband do?] He works for the village council. We have two children. [Are they going to learn Catalan?] No, although they get a little of it in elementary school. They have one hour of Catalan per week. Now they are beginning to learn English and that makes the Catalan a bit difficult.

[The third woman is originally from Barcelona.] [Where are you from?] I'm from Spain and have been here for six years. I was born in Barcelona. I speak Catalan with my family, but not with my husband who does not understand it. He is French. [Are your children learning to speak Catalan?] No, they are learning Spanish because in my family there are those who speak Catalan and those who don't so I think it's important to have one common language and that's Spanish. The children do understand Catalan a little. [Her husband works for the railroad.] [Woman four is French but is not from the region] [Why did you settle here?] I came here because of my husband's work. He is a gendarme, one of six who are stationed in the village. [How long have you been here?] For eight years. [How do you like the village?] Cerbère is great in summer but very sad in the winter. It's difficult for people who come from elsewhere to be happy here but for natives it's hard to leave. [Are there any group activities in the village that you find attractive?] There are sports associations but people say they don't really function very well. For many activities it's better to go to Banyuls or even further, but we hate the road with all its turns. Some of the road is better than it was a few years ago. The worst is the old part left between Banyuls and Cerbère. [Last year did the associations work better?] Many things have gone down the drain in the last few years. There was a club for adults in the gym but now it's closed because of the asbestos problem. The Sardana group from the school has closed down because the person who was responsible for it resigned. [Another remarks,] there is another group that meets on Mondays but it's mostly old people and they don't like new people to "invade" it. The Catalan mentality is difficult. [A different informant, the woman whose husband is a gendarme, replies:] that is more a problem of Cerbère than a Catalan problem. When I arrived here I had a great deal of difficulty integrating into the village. When I went to pick up my children in school all the mothers there would be speaking French and when they saw me they would switch to Catalan. Women of my age! Luckily my friend —— got me out of my hole or I'd still be stuck in my apartment. [Another woman agrees saying] it's characteristic of Cerbère. They don't accept the new people well. There are some new arrivals who also don't want to integrate. If a person wants to integrate they will be able to do so. [Is there a separation between the rail workers and the others?] No, there's no problem there. [Do any of you go to Portbou?] Yes, to buy food. [They agree that Portbou is livelier than Cerbère and has more festivals, even in the winter months.] We find Portbou different. When you go into a store they are more welcoming. They are much friendlier than here. They smile more than the merchants of Cerbère. Generally we go there for the Friday market. [Do any of you have worries about the future of Cerbère?] [One replies:] this year we have one hundred more people than the year before. That's because they now ask people who come to work for the railroad to live here. Many of them live in the railroad apartment building. The apartments there are not bad. [Is it difficult to buy a house here?] They are expensive. It's a seaside town. [Another adds,] that depends on the location. Land in Banyuls is cheaper than here since there is much more available there.

How Typical is Cerbère as a Case of Lost Culture?

Cerbère is at the same time a typical and untypical case study in how the French have discouraged Catalan culture within the borders of their nation-state. The village is atypical only because of the unusually high percentage of in-migration from non-Catalan speaking areas of France and the fact that Catalan culture there has been stamped out more completely than elsewhere in the region. The language, the major symbol of Catalan identity, is moribund there in spite of the fact that there are still a few Catalan speakers left in the village. Yes, there are also a few strong Catalan nationalists left in the village, but one must be beware of associating the word Catalanist with a person's true identity. In general, most of those who claim Catalan identity see it very much the way Americans of foreign origin identify emotionally with their past but are grateful for their present status as Americans. Their ties to the mother country are superficial at best and somewhat hostile at worst. France has done its job of leveling efficiently. What remains of *Catalanitat* is vestigial and harmless, more folklore than reality.

Ironically, as Josep Pla points out in the quote cited at the beginning of this chapter, because it is on the border with the *Principat*, Cerbère has been forced to turn its back on the south. This is probably the reason why the authorities responsible for improving the conditions for automobile traffic along the Ruby Coast have refused to extend the recently constructed four-lane road all the way to Cerbère. This refusal contrasts with the projected construction of an expensive tunnel between Portbou and Colera that will effectively eliminate the isolation of Portbou from its nearest neighbors to the south. Furthermore, it was Paris that refused to improve the road between Banyuls and Cerbère whereas it was the central government in Madrid that gave the green light to end the isolation of Portbou from its neighbors to the south. Additionally, as mentioned earlier, the *Generalitat* in Barcelona has provided Portbou with a good deal of aid to compensate for the loss of the economic benefits associated with being a frontier village. The disappearance of the customs bureau and its associated businesses was a strong blow to both Portbou and Cerbère. The Catalan government has paid for restoration of the Civic Center as well as the much more expensive new port, and may well contribute to the Walter Benjamin Foundation based in Portbou, a potentially important economic draw for the village.

Although some residents of Cerbère who still have family and/or historical ties to Catalunya visit Portbou for sentimental reasons, the majority of people go there to shop or for recreation. The Portbou market remains a strong attraction in comparison with its competition in Cerbère. The majority of people we spoke to in Cerbère appreciate the wider variety of stores and cafés available in the village to their south. They also praise Portbou's business people for their commercial acumen, as well as the fact that most of them speak French, a fact that is also important for French tourists from other regions of France.

Given these and other differences for most tourists and a substantial number of Cerbère's inhabitants, that village remains more of an exit from France than an entry from Spain, while Portbou continues to enjoy the advantage of being the only coastal entry point into Catalunya and Spain beyond.

Chapter Four

The People of Portbou Speak

Portbou is a village of stairs: from the railroad station down to the village they are wide and light in color, stained with sun and the shadows cast by the tall acacia trees.

Portbou is your village, yet it is at the other end of the world. It is a village that is different from all others, artificial and with a very short history. There are no farmers, people who return at the end of the day from the fields as in most other small villages; they have neither flocks or herds, nor the land to husband them. All they have is the sea, but neither is it a village of fishermen, and the Portbou church is not a fisher people's hermitage, small and white: it is rather a large structure that dominates the village, tall with a neo-gothic facade. To get there one has to climb, climb, climb.

Portbou is a village of transients. Many come from far away, hoping only to pass the two required years in order to ask for a transfer elsewhere. It is a village that faces the sea and is encircled by mountains on its three other sides. It is a new village of houses without history. Its streets are not made of paving stones. There are only stairs, leading upwards and downwards.

Portbou is a village of the middle classes. In summer its people change their clothes in the evening and stroll slowly, back and forth from the top of the rambla to the embarkation point for boats. They wear stylish clothes, high shoes: Portbou has the air of a small city.

—Maria Mercè Roca. *La Casa Gran*. Barcelona: Columna. 1991: 11, 14–15. (Maria Mercè Roca was born in Portbou in 1958 and lived there until she was seventeen.)

We conducted thirty-eight hours of interviews in Portbou, mostly in the spring of 1997. All interviews were in Catalan. The majority of the interviewees were born in Catalunya, but some were Castilian in origin. Many were active in commerce in some way, café owners and store owners. Others worked in one capacity or another for the railroad. A few were professionals, including a person in real estate, and teachers in local schools. Some were elected officials who also ran their own businesses. A few were retired individuals, but still active in cultural and/or political affairs. Our samples in Portbou, as in Cerbère, are neither random nor stratified. The interviews are based rather on opportunity. I believe, nevertheless, they are quite representative of Portbou's active population of both Catalan and Castilian origin. If we include couples that were interviewed together, the number of recorded conversations totaled thirty-three. One of these interviews will be saved for the coming chapter on language. Additionally, in 1999, we hired a young man of the village who was bilingual in Catalan and Castilian to conduct a series of questionnaires with a small sample of villagers, some Catalan, some in Castilian. None of these individuals were in the original interview sample. Unfortunately, because of sickness our assistant was able to complete only eighteen of these interviews. Recently, in 2004, another Portbou resident assisted us by conducting an additional twenty-two questionnaires. A summary of the most important information from these two sets of questionnaires will be presented at the end of this chapter.

Since our initial research and the writing of this book we have kept up continual contact with our interviewees and taken notes of these more informal conversations. Here I present selections from fourteen of these interviews that have been chosen as representative of the entire sample. In each case I have shortened the interviews to focus primarily on the following topics: the origin of the interviewees' families (Catalan or Castilian); the language of their parental household and their current household; the nature of their personal identity (Catalan, Spanish, both, or neither); and the informants' characterization of the people of Cerbère, and of Portbou, itself.

Interviewee 1. Woman, in her thirties, born in Portbou and lived there all her life. Her father worked in the customs bureau. She is fluent in both Castilian and Catalan, but Catalan is her mother tongue.

[What language do your children speak?] Little by little they speak less Castilian, but it's still a mixture. When they play with their friends it is primarily in Castilian, but this is now changing. [Does your twelve-year-old speak with friends in Castilian or Catalan?] If the child is a Castilian speaker she responds in Castilian. Sometimes the children on their own will speak in both languages switching between them. [What is your opinion of the state of bilingualism in the village?] I am bothered by the problem. It's going badly. We Catalans are all bilingual. The others are not. [If a person speaks to you in Castilian how do you respond?] I respond in Castilian, but if I know that the person's first language is Catalan, and yet speaks to me in Castilian, it's stupid and I respond in

Catalan. Some of the people here are very reticent about speaking in Catalan. [Why?] [She responds in English.] Because they are very proud! [She continues in Catalan.] Catalans are also proud people but are more open. We are a region of transients. [What does your husband do?] He works for Transfesa in Cerbère. He speaks French at work. [Do you have relations with others in Cerbère?] Sure, we know many couples and socialize with them. We are all in the same age group. We enjoy the Catalan boat festival Cerbère has in the summer. We also enjoy having drinks with them. We often speak to them in Catalan and they respond in French. [How would you compare the two villages?] I don't know, perhaps it's a cliché that people who are neighbors are not friends. I can't answer in general. The two villages have different festivals. Our meal hours are also very different. Cerbère does not have a *Festival Major* in summer and ours is very important. [Do you feel that the infrastructure for tourism here is adequate?] No! [She then notes how difficult it is to get to other villages in the region on both sides of the frontier.] We have a terrible road out of town to the south. It's infernal, yet some say it's touristic. It certainly does keep the village tranquil. [What do you consider your nationality to be?] I would say that I am Spanish. I'm not for Catalan independence. I am also a Catalan, but Catalunya is part of the Iberian Peninsula. [Are Catalans different from other Spaniards?] That's a difficult question. Some say that Catalans are a very reserved people. People from the south of Spain are more open. They have a different spirit. But these are all clichés.

Interviewee 2. Woman, in her mid- to late thirties, married with two children. She is an elementary school teacher. She was born in Girona (Catalunya). Her parents are Castilian speakers from the south of Spain. They now live in Portbou. Her father came to the village to work for the railroad. Her home language was Castilian. During her schooling up to the age of twelve all her classes were in Castilian.

[Did you find it difficult to learn Catalan?] Yes, it was a great effort for me. My parents still do not speak Catalan and they feel the need to do so. I now speak Catalan all the time, even with my parents who do understand it, but respond in Castilian. My sister, who is nine years younger than I, speaks Catalan very well. It is natural for me to speak in Catalan with my children. [Does the fact that children are in a Catalan school influence the language they speak at home?] In the long term yes, in the short term, no. [When kids play in the schoolyard what language do they use?] Until recently most of them spoke in Castilian. Now you hear both languages. [What do you consider the future of Portbou to be?] It's very difficult. People who have a bit of money are more or less satisfied. It's quite closed here. [How would you describe the people of the village?] They are difficult. [Do you think that the large number of retired people affect the spirit?] Yes, it's not a very lively village. [When asked what your nationality is by a foreigner what is your reply?] *Catalana*! [Do you feel that Catalans here who have deep roots in the culture accept others as Catalans?] Yes, they are considered to be Catalans if they speak the language. [When Castilian people attempt to speak Catalan does that give you pleasure?] Yes! There are many who do not wish to learn to speak the language. [Do you believe that people who are Castilian speakers and have children and grandchildren who go to school here will lose Castilian?] No, never! That's impossible. There is the TV. [But, we say, there are Castilians who do feel that they will lose their language in time. What do you think of that?] That's a political war. No, it's impossible. [How would you describe the average Catalan?] They are a bit cold. Those I know think before they respond. They are very reflective people and work hard. [What do you think about people who have lived here for a long time and do not speak Catalan?] For our parents who came here during difficult times and had to earn their living the situation for them was very hard. It's difficult for them to change their language. Also there is no necessity, economic or cultural, to learn Catalan. [So do you feel that it is difficult for an adult to learn Catalan?] No. I know a Bosnian who learned Catalan and speaks it very well. It's certainly not difficult. [She adds] Catalans are nonviolent.

Interviewee 3. Woman, thirty-two, married with children. Born in Portbou. Mother is from Portbou and father is from Madrid. Her home language was Castilian. She has her own business while her husband works for the railroad. Her husband's family language is Castilian.

[What language do you normally speak in the house?] Castilian. [What language do you usually speak with your friends?] Catalan. For my husband it depends. With his old friends he normally speaks Castilian. [This is a common practice in Portbou, where people tend to use the language they first spoke with specific friends.] [What language did you use in school?] Castilian. At school now everything is in Catalan. [Do you know what language the school children speak when they are at recess?] There are children who play in Castilian. My nephew speaks Castilian with his friends in school. [Do you have relations with people in Cerbère?] Yes, I have relationships with people in Cerbère and also in Banyuls. [What language do you use with these individuals?] I don't really go to Cerbère often now, but when I do I speak either Catalan or French there. [If a person asks you to identify yourself what do you say?] I say I am *Catalana-Espanyola*. [Are festivals important here in Portbou?] Yes, they are very important, but only in the summer. They are an important source of activity. There are also several associations in the village. [If a person is not born here, as you were, but works here and does not seem to enjoy the idea of learning Catalan, do you feel that people in general tend to consider such a person as a Catalan or as a stranger?] Always as a stranger. The people here who consider themselves Catalans are very closed in. Those from elsewhere are always considered to be different. [What about the outsiders who do learn Catalan?] They are somewhat accepted but they are never accepted as natives here.

Interviewee 4. Man, thirty-five years old, single. Born in Portbou. Parents lived in the northwest of Spain, but father was from Aragon and mother's parents now live in Portbou. They immigrated to Portbou to find work. He has two sisters; one lives in Madrid and the other in Figueres [a small city not far from Portbou].

[What language do you speak with your sisters?] We always speak in Castilian. It is our parental language but both my father and mother speak Catalan well. [When do your parents speak Catalan?] When they go out in the village. [What is your education?] I have a college degree. [Why have you stayed in Portbou?] I'm not really sure, it's mostly a question of circumstances. There is not much work here and many of the unemployed have the same training that makes it difficult to find work. [Do you feel that you are speaking a foreign language when you speak Catalan?] I'm not sure. Both languages are very interiorized. Even though Castilian is my first language I don't have the feeling that I'm speaking a foreign language when I speak Catalan. If a difference exists between the two it's very small. Sometimes clichés in Castilian come to mind when I'm speaking Catalan although when it comes to numbers I think in Catalan. My schooling was also in Catalan, in a private, religious school. [If you are having a conversation in Catalan and a person arrives who speaks Castilian what do you do?] I ask if the person understands Catalan and if they do I continue in Catalan. If I speak directly to that person I will do that in Castilian. [Do you have any difficulties switching?] No, none at all. I don't pay any particular attention. [Do you go to Cerbère?] Rarely! I do go to Banyuls, particularly for night-time activities. [Do people from Cerbère frequently come here?] Yes, and for many reasons. To the gym, for example, and to buy things. [If you were the mayor of Portbou what priorities would you have to improve the village?] Three things: First, find out how people in Portbou can work to improve relationships with the north. [He means the north of Spain.] Culture is the most important thing for the future image of Portbou. I don't mean museums and music. I have a wider image. Culture has to be put in the context of leisure. Here in Portbou they can create a small culture industry. Beginning with the Benjamin Foundation and Dani Karavan [the architect of the Benjamin monument]. Portbou needs to become a destination rather than just a place to pass through. Then come the questions of the retired people, the road problem, and the railroad. Some culture already exists in Portbou. This can be extended beyond Benjamin to other activities. [Does Catalan culture play a role in all of this?] Catalan gives it a more local dimension. It forms part of the whole. [Are there people here who are interested in culture in the way that you are?] Yes. Most of the young people are not interested, however. Football (soccer) is very important here. The football club, *penya*, is more than sixty years old and is one of the oldest in all of Catalunya. There are many members here in the village—older men but also women and children. They get together to watch the games. [Is the election of the *hereu* and the *pubilla* an important event?] [This is a yearly event to elect a young man and a young woman to represent the village based on their personalities, their interests in improving the village, and their capabilities. The vote takes place during a meeting of the town council. Every year in every town, region, and, finally, throughout the nation, the hereu, and the pubilla are elected by the appropriate political figures.] No, in my opinion it is "folkloric" in the worst sense of the term. [What about the Canigó flame on the twenty-third of June?] It's not very interesting to me. I don't like fire and fireworks. [Is Portbou Catalan or Castilian?] Portbou is a microcosm. [What do you mean by that? Is it a little Barcelona?] Yes, but it has a wide variety of people. It is a frontier town. It is also between mountains. [Are the people of Cerbère Catalans?] No, they are French. [How would you describe Catalans?] There are many clichés. There are also clichés about people from other regions of Spain, such as Andalusians and Basques. There is a certain envy of Catalans because Catalunya is not naturally rich but has managed to make itself rich. It has also given many important people to the world. And it has its own language—a language that one normally uses everywhere. That's something people tend not to realize when they come here for the first time. [Do you feel that trial judges should be able to speak in Catalan?] [A new law passed in 1997 not only required that a person has the right to have his/her trial take place in either Catalan or Castilian, but also that judges have the right to use interpreters in such cases. Thus they are not required to actually speak Catalan and many court judges come from non-Catalan-speaking parts of Spain. Some Catalanists would require that all judges be bilingual.] It should be optional. Many judges come from the outside. [How do you identify yourself, as Catalan or Castilian?] I am without a country. I am a citizen of the world. I don't like frontiers. I am for Barcelona (football) when they play Madrid, but if Madrid plays well I am glad of it. [Do you feel that Catalunya should have its own selections for the Olympic Games?] Yes, why not! [Do you feel that it makes sense to study Catalan identity?] Yes, it does. But it's very complex here. Many of the people in Portbou are here only temporarily. [Are there associations in the village, and if so, how do they function?] Most of the groups are small. [Do you have the impression that there are two communities here, one Catalan and the other Castilian?] Sometimes I do think so. There are Catalans who have lived their entire lives here and they do not seem to care at all about new residents. There are old folks here who have little or no contact with the outside or with outsiders.

Interviewee 5. Man, in his late sixties, born in Portbou. Has spent most of his life in the village. Lived in Barcelona when he first worked for the railroad, but then transferred to Portbou until his retirement. During the Civil War he went over the border into France for some time.

[Can you tell us about your family?] My parents were Castilian, from Burgos. Father came here to find work. He worked for the customs agency. Before it was quite easy to integrate into Catalunya because there was so much less immigration. It was also easier to learn Catalan for the same reason. I don't make any distinction between Catalans and Castilians. One language should not be imposed on another as it was under Franco. The initial years [under Franco] were very difficult for my father. Politically he was on the left. He lost his job but found another, but nothing stable. After seven or eight years he was able to rejoin the customs bureau. [How did you learn Catalan?] Under the Republic both languages were taught in the schools. I took adult courses later to learn to read and write. The children here learn Catalan well, but outside of school they tend to speak in Castilian. They often play in Castilian. During the war the school here was closed and for a while I took courses in Cerbère—in Castilian for us! I liked going to Cerbère. I don't like frontiers. We were in school there for about six to seven months. The people of both villages knew one another well. [What kind of relations between the two villages are there now?] Minimal ones. Mostly touristic. People from Cerbère do come here for Catalan classes. [Do you know the history of the monument in Cerbère for the dead from the First World War?] Yes, many people from Portbou went to the other side to fight against the Germans. Many people on both sides of the border have Catalan family names. [When you are in Cerbère what language do you speak?] When I go to France, that is to say, Northern Catalunya, I speak Catalan. All the old people in Cerbère speak Catalan. Catalan is now marginalized in France yet the gypsies in Perpignan speak it as their first language. The French see the Catalan language as inferior to French. Catalan is spoken in France only in the mountains. [Do you have the impression that people here are proud of their language?] Yes, after all, your language is your identity and your culture. [What do you think are positive and negative factors in Portbou?] Portbou is in a major crisis due to the disappearance of the border. It is evident that Portbou's economy depended on its location and the railroad. In the past it had an excellent economy. I don't think the village can live on tourism. It needs something else to survive now that the frontier no longer provides employment. It needs some kind of industry. Many French do come here to shop. For now things are cheaper than in France, though alcohol is really the only saving for the French at present. There are very few places here for people to spend their vacations and few houses for sale for retired people. The new port is certainly not a panacea. The road between Portbou and Colera is very poor, but a project does exist to improve that. There is no solution for real continuity between ourselves and France. I don't think a better road between Port Vendres, Cerbère, and here will be built. Portbou must invest a great deal of money in order to prepare for tourism. Colera has much more land on which to build than we do here. Investors are interested in putting their money into a larger town than Portbou with its limited opportunities. They would not want to build a small hotel with just a few rooms. However, some things have been accomplished here: the port and the maritime passage. Also the Civic Center is quite interesting. If the planned sports complex were to be built it would help also. It's all very difficult because the village gets little in taxes, beyond what is needed for necessary services. The railroad company does also provide some support. [What about culture?] The priest here is very important for the village. He participates in the life of the community. He has frequent meetings with married couples. [How do you declare your identity if someone asks?] I am Spanish. It's clear that I am also Catalan. But for me I think we have to get rid of all these questions. I am Catalan and Spanish. [If you speak in Catalan to a person and he or she answers in Castilian what do you usually do?] If I realize that the person does not speak Catalan I will answer in Castilian. If he does not want to understand that is another question. I'll continue speaking in Catalan. Portbou is very different from elsewhere. Perhaps the majority in the village are Castilian speakers. There are those who just do not want to speak Catalan. People who are from the working class integrate better in Portbou than people who are functionaries. Many functionaries just do not want to integrate. If you oblige people to conform they will not want to, but there are those who do want to. The [former] mayor does not speak Catalan well. I speak to him in Catalan but he has difficulty in understanding what I say. He can explain himself better in Castilian. These are difficult questions. Some people feel that they are better than others. We are all equal. I'm not better than someone else because I'm Catalan. [Does Catalan have great value for many people?] Yes, Catalan was prohibited many times in our history. Many Catalans were for the Republic and against Franco. The *Generalitat* wanted a Republic here in 1936. [Do you think that the younger generation here will still value Catalan?] It's not a question of being practical, it's cultural. If you lose your language you lose your culture. [Do you think that this is important for the young?] It's complicated. There are all kinds [of people]. There are those who value the language and others who don't. The new generation doesn't know about Franco. It's possible that this knowledge will be lost. The 1983 language law was modified in 1997. Some said it was not necessary to do this. It's a very complicated problem. There are some Catalan intellectuals who are convinced that the language will be lost. It's difficult to analyze all this. Politicians will often exploit issues like this one. To keep a language alive people have to want it.

Interviewee 6. Man, in his forties, married, with two daughters. He is from a non-Catalan province. He is a businessman in the village and works with his wife.

My wife [born in Portbou in a non-Catalan family] and I speak Castilian with one another, but my wife speaks Catalan with our daughters. [How did you learn Catalan?] I learned it in the street. I came here in 1992 and, at first, worked for the customs bureau. My wife and I realized that work there would end, and so we opened this shop nine years ago. Work at the customs bureau began to fall off when most of the traffic shifted to trucks and passed the frontier at La Jonquera [the border town on the French-Spanish autoroute] [Will you stay in Portbou when you retire?] Probably. My wife is from here and I have made my life here. [Are there commercial relations between Portbou and Cerbère?] Some people from each village are friends but we have no commercial relations with them. [When you go to Cerbère what language do you speak?] I speak Catalan. Cerbère is in Northern Catalunya. They speak Catalan differently from here. [How would you compare the two villages?] Portbou is larger and has more vitality than Cerbère. Many people from there do their shopping here. There's more work here for the railroad and commerce. [How do you see the future of Portbou?] I'm for the port. It's another form of life for the village.

People from all over will come here. Myself, I don't have a boat because summer is the high season for our business. The tourist season begins to fall off at the end of August but some retired people continue to visit in fall. [What do you think of the Benjamin monument?] I don't like it. They should have built something more beautiful, a little garden, for example. [Do you think it attracts people to the village?] Not many. There are Germans who do come here to visit it. There are other people who should have been honored who are more famous than Benjamin was and who are Spanish. [Don't you think that Benjamin is well known?] I don't think so. We know that his work is very difficult to understand. It's probably better known by people who understand history. [Do you speak Catalan or Castilian with the people who come here to shop?] Depends on what language they speak to me. I don't speak much French. When I speak Catalan I am translating from Castilian in my head. I don't speak Catalan perfectly. [However, he seems to speak it with ease.] [Do your children make fun of you when you make mistakes?] No, not at all. I took a course a few years ago in order to read and write the language. [Do you have the impression that you live in a Catalan community or a Spanish one?] Both of them. All this is Spain. It's a complex thing. There are people who have a different idea of this. We learned in school that Spain is all of these regions. The old people here speak in Catalan. [Do you think that the language will disappear one day?] In school, courses are taught in Catalan. There are many people who work here for a short time and they feel it is not normal for their children to be educated in Catalan. It's a problem for people who come here to work. It's clear that Catalans are bilingual. The two languages are co-official.

Interviewee 7. Woman and her husband, in their mid-fifties, who, together, operate a food and liquor store in the village. They were interviewed in 1998 rather than 1997. She was born in Girona and her husband lived with his parents in Portbou.

[What was the language used here when Franco was in power?] [Woman] Most people spoke Catalan. But in Girona they spoke it even more than here. There were many Civil Guards and railroad workers here even then and most of them did not speak the language. There are now many people in Portbou who can write in Catalan, yet they tend to speak Castilian because it's the family custom. Our daughter, however, speaks Castilian with her husband, but her children are Catalan speakers. The children speak Castilian with their father and Catalan with their mother. [And with their friends?] Often they speak in Catalan. [Where are most of your friends from?] Most are from the village, and even those who are not from here generally speak Catalan. Every one of them from elsewhere speaks in Catalan except when they are angry. [If a person is from elsewhere but speaks Catalan what do you consider them?] [Emphatic] If they live in Catalunya and speak the language they are definitely Catalan. Yes. No one cares what his or her maternal language was. Now, however, when most of the young people come into the shop they say "Buenos dias." We respond, "Bon dia." They all understand Catalan. [She adds that it's normal that when people go to France or Germany they have to learn the language there and that in Catalunya it's not required of them.] [How are relations among businesspeople here?] Not very good. There is a great deal of envy here among them. [What is your opinion of the Civic Center?] There should be more activities there. Most of the men go there to play cards and the women to sew. There does not seem to be much interest in doing things with the young people of the village. [Is this town much different from most others?] Yes! There is not enough collaboration here among people. In the past it was easy to make lots of money but now it's very difficult and cooperation is necessary to survive. We hope that the port will be a solution for many of the problems in Portbou. We also need to improve the road to Colera but that depends upon the central government in Madrid. We have no right here to touch it. [What do you think is the future for Portbou?] Many think it's very bleak. We don't see it that way. We have hopes for the village. While it's true that most young people don't come here, that could change. Port de la Selva has many visitors. Portbou could develop itself in the same direction. There are nature and historic trails that could be developed. [They also note that most of the food in the restaurants is not of the highest quality and that's bad for tourism.] There is a possible project to mark the trail that Walter Benjamin took to get over the mountains from Banyuls to Portbou. Many people here do not really know about Benjamin but that can be changed. [Next to commerce what are the other economic activities here?] There is also honey production [one man] and the railroad, but the latter is run from Madrid. There is another problem for the economy. People who vacation here in summer do not pay for garbage collection. It's free for them and not for us. That puts a drain on the village. [In general how well have Castilian outsiders integrated into the community?] There are many who have integrated well but others who have not. When one is in business here one has to speak Castilian. With Castilian clients I always speak a bit of Catalan to them. I think that's very important. They usually do understand me. I will repeat what they ask for in Catalan if their request is made in Castilian. [Why do you do this?] Because I want them to know that they are in Catalunya. One of the men who has a café on the beach speaks Catalan perfectly but does not speak it with his family or with most of his clients who are Castilian. In our social group we speak in Catalan and the Castilians understand us.

Interviewee 8. Man in his sixties, married, lives with his wife. Now retired. Formerly worked at the money exchange in the station. He was born in an apartment of the Casa Gran, the residence for railroad workers, as his father worked for the railroad.

[What language do you speak best?] I speak Castilian better than Catalan because that is the language in which I was educated. My father was not Catalan. He was born in the north of Spain. He came to Portbou when he was about two or three because his father had found work in the customs bureau. My family arrived in around 1920. All my brothers were born in the village. [What was the language of your household when you were a child?] Castilian. Now we all speak Catalan. My mother is Catalan and comes from a district not far from Barcelona. Both my paternal and maternal grandparents spoke in Castilian.

[Is there much work for the young in Portbou?] No, not now, but during the Second World War there was a great deal of work in the village. After the war Franco declared Portbou a protected area. During the Civil War the fascists bombarded the village from both the air and the sea. Now that the frontier is open there's less work. The railroad hires very few youths from the village. Many who live here work for Transfesa in Cerbère. I don't see any small industry coming here. Portbou is no longer in its golden age. There is much more work in Cerbère. [Do you go to Cerbère often?] No.

[Is Portbou a Catalan town or something else?] It's in Catalunya, but if you take the number of Castilians living here you might think it's a town in Andalusia. [Do you speak Catalan in the street when you meet people?] I, yes! When I was fifty I took my Catalan first name back. Before that I was obliged to use the Castilian version of it that's still on my identity card. At the same time I decided to speak Catalan most of the time since I am a Catalan even if I do speak the language badly. I also write it badly. [If a French person asks you your nationality what would your reply be?] I am a Catalan from Spain. [When a person speaks to you in Castilian what do you do?] If I know that the person lives here I respond to him in Catalan. A person passing through, perhaps someone from South America, I will speak Castilian with him. To be courteous to him, but not as an obligation, I will speak Castilian with him.

[Is your life here pleasant?] It's tranquil. Most people come through Portbou by Talgo [a relatively new and fast express service that does stop in the village, but most passengers are in a hurry to go elsewhere] and do not get off in the village. For me that's good. If they do stop it's for a coffee or a meal and that's all. [Do you feel that there are enough activities in the village?] Yes, there are. There are many bars. I sing in the chorus, and I have several hobbies. [Is it easy to buy an apartment here?] Yes, but they are very expensive. [Do you think that more secondary residences in the village would be helpful?] Yes, there are many empty houses in the middle of the village. Many people from Portbou who have retired to Figueres still own houses, but are rarely, if ever, seen. Usually people who don't live here permanently and come for vacation stay in the houses of their parents or have inherited them after their parents' deaths. That, in fact, is the rule.

Interviewee 9. Man, in his thirties, not married, works as a professional in Portbou. Was born in the province of Girona. Parents are Catalans. Came to Portbou in 1991.

[What is your impression of Portbou?] It is a very closed-in town. We don't use the train much and the road out is very difficult. The social situation here is very special. Portbou is going through a very difficult period. In the past there was a great deal of money here, but it was a false situation. No local industry or agriculture was created. There is now a feeling of nostalgia. And the younger folks have to leave to find work. Additionally, there are many transients here who stay only for a short time. The police, for example. There are many retired people, and there is a sentiment of sadness. There are also people living here as if they were living in Spain. The sense of being Catalan is superficial—it's mostly folkloric. The people live here as they could anywhere else. They have little esteem for the town. In general they don't implicate themselves in issues that effect the village. They haven't created a strong sense of community. There's a great deal of divergence here among the people. [Do most people speak Castilian or Catalan?] I always speak in Catalan. I do have problems sometimes speaking Catalan with some people who aren't Catalan speakers. In general, however, when I'm with a person who doesn't speak Catalan I speak Castilian with them. But with people who have been here for a long time I'll try to speak Catalan if they understand it. There are also conversations in two languages in which one person speaks Catalan and the other Castilian. People do have a tendency to switch from Catalan to Castilian, but I don't do this. I prefer to speak Catalan and most people do realize this. [Why do you persist in speaking Catalan?] Because it is my right, and this is a country that was occupied by an army. We should not forget that! This is historically a nation and we should remember that! There are many people here who do not know the history of this country. [What do you see as the future of Portbou?] Very difficult. The port is not a solution and is very expensive. [What alternatives would you suggest?] I think that at this moment Portbou should exploit its particular potentialities. There should be more cultural activities as an attempt to better integrate people into the community. The Civic Center is a positive development but it's not used enough. The Benjamin monument is a wonderful work of art, but it does not really have a function for the community. Also it's been a kind of disaster. You have seen the broken glass! The town is closed in. People go up to the graveyard to bring flowers for their dead relatives, but they aren't really interested in the monument right next door. Benjamin is not well known in this country. He does not attract tourists here the way Dalí might. [If someone asks you your nationality what is your reply?] Catalan. I am disappointed in the direction this country is going. The response of people having their language forbidden is either one of revolt or passiveness. Theoretically, we have the right to speak Catalan but psychologically many people still feel inferior about speaking it. It is easier to change languages than to explain the history of this country. This is a society that is very conformist. The majority opinion here in Portbou is not sad or pessimist; it's realist! People here often diagnose the situation correctly but have no idea of how to act to change the situation. Perhaps the easy life that was led here in the past has left its mark. The people do not profit from their environment. There are no farmers here to exploit a possible revival of grape agriculture. As for tourism, it has entered into a permanent crisis in the last few years. People no longer take a month's vacation. They spend a few days here and a few days there. [Do you enjoy visiting France?] No, I don't like France. When I find a person who speaks Catalan I will speak with them in Catalan. We find it very agreeable to have you here learning the language. It's a pleasant surprise for us. Most outsiders speak only Castilian. [What is the attitude in Portbou about inviting people to their houses?] It's generally not done—it's a Catalan characteristic. There's a clan structure here and most social relations are very superficial. In Spain everyone goes to the bar. Here people tend to go home.

Interviewee 10. Man in early forties, was born in Portbou of Catalan parents. The home language is Catalan. He also speaks Castilian and French and learned the latter in school. He learned Castilian in school where it was the normal language under Franco. The first time he was exposed to Catalan in school was in Llancà at the age of fourteen and found writing in the language very difficult.

[Do you have social relations with any people of Cerbère?] Yes, there is a choral group there and that's one of my interests. [What are the relations with the Civil Guards who are stationed here?] Before we had local people who served as local police. We had good relations with them. When the Civil Guards came we didn't like them. Most of them were [and still are] Spanish and not Catalan. Many of them didn't want their children to learn Catalan and most continue to live in their barracks. It's like a ghetto. The National Police here are not required to live in barracks. They buy or rent in the village. Many of them are also from outside of Catalunya but they have become better integrated into the village than members of the Civil Guard. Some of the National Police have married people from Portbou. [What is the history of Portbou's economy?] Portbou was originally a fishermen's town but it's really a product of the railroad. In the 1960s, truck traffic on the new toll road through La Jonquera began to cut into commercial traffic here and an economic decline began. Portbou doesn't have a sufficient touristic infrastructure. When things here were going well the mayors didn't think that tourism was necessary. Unlike Colera and Llancà, there was no investment in tourism. Now I think it's too late for the small population that remains. It seems to me that the maximum that could be supported is around a thousand. Now they have begun to construct the port and there's a plan to ameliorate the road between here and Colera. They have also built the new school. They may also build a new sports complex but I feel that all these efforts come too late to help. There is little work here and many people have to go to Llancà and Figueres to find work. Also there are too many retired people in the village. The birth rate is very low and there are few intra-village marriages. There are those who commute to work but in the long run they tend to move elsewhere. There is very little space in the village for new construction, and land is very expensive. The price of construction is very high due to the terrain. It is much cheaper to build a house in Llancà, for example. Many of the young people from here buy apartments in Llancà. The local population here is very closed in on itself. Most are not interested in cooperating in communal activities. Those activities that do exist depend on a small group. Every year it's the same people. The nucleus is made up of members of the *penya* [the fans of the Barcelona football club] the choral group, and the town council. [What do you think of the Civic Center?] It's a very beautiful restoration but there are very few activities. There's no one to act as a stimulus. The library is underutilized. [Are there any other activities?] There's the *Full* [a magazine edited and published by villagers] and the priest is very active. [What is your opinion of the new language law?] I am a Catalan nationalist. I consider that Catalunya is a nation and that we should have more rights than we do at present. We should attain these rights little by little and without violence. I am even for independence within the Spanish community. [Do you think that Catalunya should even become independent of Spain?] That depends on how the European Community deals with autonomous regions. It's a complicated problem. [What is your impression when you are speaking in Catalan to someone and they respond in Castilian?] A few years ago when someone spoke in Castilian everyone changed to that language. Now this is less the case. But it still depends on the neighborhood and with the passage of time it gets better. The people are now more accustomed to speaking in Catalan. In fact, many are becoming used to speaking both languages. If you ask people from the outside who speak Catalan perfectly if they are Catalan, they might surprise you by responding, "No, we are from Andalusia." There are also conversations in both languages with one person speaking Catalan and the other Castilian. It's a question of economics. Castilian is important for commerce. [If someone asks you your identity what do you reply?] I am Catalan. My nationality is Catalan. To speak Catalan is an important sign of identity. Two cultures can fuse but it's more difficult for two languages. Language is a symbol of the identity of a people. Here's the problem: all the children now study Catalan in school from the beginning. But it's clear that it has lost some of its value. It's normal for children who speak Castilian at home to speak it because it is the language of their family. But this means that Catalan loses some of its value. Such children might not value Catalan as a language; it can carry an inferior value for them.

Interviewee 11. Man, married, in thirties, with two children, born in Portbou. He works for the railroad.

[Do you feel more Catalan, Castilian, or both?] I am Catalan. My circle of friends is also as strongly Catalanist as I am. We are all Catalans. [Are there many young people in your age group of like mind here?] Yes—well, not many. When we talk about identity there are very few at present. When I'm at work I speak in Catalan with everyone. Those colleagues who don't speak it do understand it and Catalan is the language I always use. [Where were you born and what is your family background?] I'm from Portbou. My mother is from the *comarca* [county] of Empordà and my father is from this village. My grandparents were also Catalans. [Are there many people in Portbou that have their roots in the village?] Yes, there are, but it is also a village with a great many people who don't stay here. There are transients. [How do you see the future of Portbou?] It will become a dormitory town. Many people will live here but work in Figueres. [Is there adequate housing here?] I think that there's enough housing for the people who are natives, but not for people from Figueres, for example. [What is your opinion of the new port?] It's very bad for the local ecology and I'm against it. I think the best solution for the village would be to exploit the natural environment as such. [What is your opinion of the Benjamin monument?] The people here know about it as a possible attraction for tourists. I hope that more activities take place here that have to do with Benjamin's work, but I am afraid that most of these activities will take place in Barcelona. [What is your opinion of the cultural life in the village?] Culture's not very important here now. Before, in the old days, there was more interest. There were also more contacts among the people of the village; for example: visits to sick people and more interaction among the young. Now there is too much individualism. [He notes that the priest is important for the Christian community of which he is not a member.] [Are there relations between the Castilian and Catalan communities here?] Yes, due to the fact that the Catalans are "soft" in their relations. If one speaks Castilian to a Catalan it is the Catalan who has to give in and switch languages. I have no relations with these people. There are many people who have lived here their whole lives and who do not speak Catalan. But for most Catalans it's too easy to switch to Castilian. [Do the children, when they play at school, use Catalan or do they use Castilian?] It depends. The majority of them play in Castilian, but there are some small groups that do play in Catalan. It's easier for the young children. [Do you feel that since there are courses in both languages that one language will come to dominate over the other?] I don't know. It's a question that has to be posed. It's obvious that Castilian will not disappear. [Sonia notes that one mother told her that she speaks Castilian with her children because they get so much Catalan exposure in school. She adds that some people think that the children get *too* much Catalan in school.] There are many different attitudes here. It's true that many of the children here will leave Catalunya and will need to speak Castilian if they move to another region. It should be the rule that all Spaniards learn the three other national languages of the country. I consider the new language law here too weak.

Interviewee 12. Woman, about fifty years of age, widowed with two sons. Her family is from elsewhere in Spain and spoke another local language in the household as well as Castilian. Father was a member of the Civil Guard stationed in Portbou.

[When did you begin to speak Catalan?] I began to learn at the age of ten. [She now speaks and reads Catalan well but does not write it perfectly.] I began to speak Catalan seriously when I began dating because I went out with a Catalan boy and married a Catalan. [In her husband's house everyone spoke Catalan.] My husband's family are all Catalans. *My* house was another world. I remember the first time I heard Catalan in school. The teacher said to the student: "Speak Christian." We lived a tranquil life in our household because we did not know anything else. I later met some Catalanists in Girona and I learned that people should defend their language and their customs. We began to understand more. My education, however, is very limited. I am an autodidact. Like the young people here at the time, I went to work immediately. I got a job at the customs agency. Money at that time was easy to find. Everyone worked. I began in 1962. [What about your mother?] At first she ended up in a town where no one spoke Castilian. Everyone was a Catalan speaker. It was very strange for her since at that time no one tried to learn it. At the same time my father in his very Castilian environment never learned to speak it. [Today do the Civil Guards still speak Castilian among themselves?] Yes, they have good reason since they are viewed as an occupation force. People here didn't want to have any relations with the Civil Guard. Some families like mine did eventually integrate but most of them did not. [What role does the priest play in the life of the village?] He's a wonderful person and very open. After each of his monthly discussions, which everyone is free to attend, the entire group goes off to a restaurant (a different one each time) to eat together. Do you know any very militant Catalanists in the village? Yes, there is one ——. [She then begins to talk about the festival of Sant Joan and the lighting of the flame on the Canigó Mountain in French Catalunya.] The Canigó is the summit of Catalanism. The flame begins there every year and is then taken south, even to the Balearic Islands. All of Catalunya lights the bonfire at the same time of night, at 8:00 p.m. There is always a doll on the pyre. And the pubilla is always present. [Is there participation of Cerbère in this custom?] Yes, some people from Cerbère come with the flame and then Portbou transports it to Colera. [When we observed the ceremony in Portbou the flame was brought by someone from Collioure along with a message about Catalan identity. Informants in Cerbère, on the other hand, told us that while they did celebrate the event as the beginning of summer, for them its significance went no further.] [What do you know of the exchange program between the two village schools?] The two mayors were very euphoric. They created interchanges between eight linked villages, four in France and four here. These were good ideas but politics are politics. I know that the French government was against the school project. The French government is very totalitarian. [Are you a Catalanist?] I consider myself more as a person of the world although I do esteem Catalunya. I am not a *Catalanista*. I am not an extremist. [She then notes that both her mother and her sons all speak Catalan.] I speak Catalan with my sons. With my best friend who is also of Castilian origin and who speaks Catalan well I speak Castilian because that is how our relationship began. We went to school together and Castilian was always used there. [How do you identify yourself to another person?] I would say I am Spanish. But it depends on who is asking the question. If someone in France asks, then I say, I am Spanish. But if Catalunya were to become independent I would choose it. I do feel Catalan. [When a mixed group gets together what language is usually spoken?] Often I go out with fifteen or sixteen people who are all friends. If I'm next to a Catalan person I will speak Catalan with them. If the person has the habit of speaking Castilian, I will speak Castilian. For us it is not a traumatic situation. For some who are Catalinists they will not accept the fact that you speak Castilian. They will say, "You speak Catalan, so we will speak Catalan." [Do people here sometimes mix the two languages?] There are some "barbarisms." When I write a letter in Catalan I show it to our priest and he says it's poor Catalan. But I have my way of writing it. [Do children correct their parents' Catalan?] My older son often corrects me. [Do you have some ideas about how Portbou could improve the current situation?] Yes, I am very much in favor of the Walter Benjamin Foundation, but it's very difficult. There are so many foundations in Spain and most of them don't last. Also it's difficult to organize cooperation between Spain and Germany. It's a question of finding money and competent people to handle it. [What about the new port?] I think that it is a bit too large for the village. But if you want to come back here after the summer months you will not return without the port. Before, everyone made their living either from the railroad or the customs. The young here now have very little access to the railroad for work. For me the best type of tourism in Portbou would be secondary residences. In any case Portbou needs something to get out of its difficulties. [What is your opinion of Cerbère?] It's rather negative. Some intermarriages have occurred between the two villages and for those who stay in France it has turned out to be very difficult to adapt to the life there. Perhaps the Spanish are more open than the French.

Interviewee 13. Woman, in her thirties, lives with her husband in the village and does part-time work and, has two children. She is from outside of the Catalan-speaking area of Spain. She was one-year old when she arrived in Catalunya with her parents, but she lived only for a short time with them because they died when she was young. She attended boarding school from the age of seven.

[How come you speak Catalan?] It is very difficult for me to speak it, but I like the language and learned to speak it on my own. [How did you come to Portbou?] My older sister often brought me here for vacations. [What does your husband do?] He works for the railroad. [What language does he use?] He understands Catalan, but speaks Castilian. We speak Castilian in the house, but the children are taught Catalan in school. [What language do you speak outside of the house?] With Catalan speakers I speak Catalan, but most of my friends are Castilian speakers. [Do you know about the interchange between the Cerbère and Portbou schools?] Yes, my daughter was in it and did not like it at all. She didn't know the teachers and had no confidence in them. [Do you ever go to Cerbère?] It's been a long time. It's not worth going there. The prices are very high. I do go to Banyuls to shop at the supermarket where it's much cheaper. The exchange rate is not good for us in France. Here we accept the French franc but no one in Cerbère accepts the peseta. [Can you compare the two villages?] Not really.

[How do you feel about Catalans who are nationalists?] There are some, "Catalan-Catalans" who are very critical of Castilians, even those who speak Catalan but make mistakes in the language. There are other Catalan speakers, *Xarnegos* [a pejorative term for people of mixed Catalan-Castilian background] like me who are more tolerant. The others say "if you don't know how to speak Catalan, don't speak it." [Can one live here speaking only Castilian?] Yes, it's basically a village of workers and Civil Guards. [How do you identify yourself?] I am Spanish but if you ask me where I was born, I say I am someone who has "taken the long road." [Are there others born in Spain and who speak Castilian with their husbands, but who also speak Catalan?] Yes, there are many. If the mother of the family is Castilian the children will speak Castilian. If the husband is Catalan the children will speak Catalan with him and Castilian with their mother. [Can you write in Catalan?] No, it's very difficult. [Do you think that because your daughter is taught in Catalan she will lose Castilian?] No, never. That would happen if Catalan were the only language here. There is Castilian radio and television.

Interviewee 14. Woman, in her late forties. Catalan mother, Castilian father. Runs a café with her husband. They have two children. She was born in Portbou. Her family came to Portbou in 1905–1906. The language at home was Castilian because of father's native language. Her husband, who is not Catalan, understands Catalan, but does not speak it very well.

[It's obvious that you are a Catalan speaker who is also bilingual] Yes, I like to speak Catalan. I speak it with my older son but he does not like to speak it and I don't know why. My younger son speaks Catalan with his friends. [Do you think that Catalan will survive in the future?] Yes. [What language do you speak with your mother now?] I speak in both Castilian and Catalan. I feel it is important to know about Catalunya. These are our roots and one's roots are very important. [How would you characterize the situation in Portbou?] It's not very good. [How would you characterize Portbou?] It's a mixed village with people from everywhere. Those from the south of Spain are very closed in. Catalans are too but when they open up, they are very open! There are all sorts of Catalans here. Some are more tolerant than others. The most intolerant are those who are not Catalan-Catalan. [The reader will note that this contradicts a statement by the previous interviewee.] For example, there is an unemployed young man here who wants to speak only in Catalan. He was finally hired by a Castilian company, but continued to speak only Catalan and was fired. His father is from Madrid and his mother from Aragó! [If you are speaking Catalan to someone and another person shows up who speaks Castilian but who also understands Catalan what happens?] I will change to Castilian, particularly if the person doesn't understand Catalan. My father didn't care if others spoke Catalan in, for example, some public meeting because he understood it. [If a person is not born a Catalan but speaks the language is he or she considered a Catalan?] There are all sorts of reactions here in Portbou. It's not a question of the language but of the person. If a person has lived here and wants to integrate there's no problem, but if someone lives here for a long time and speaks ill of Catalunya or Catalans that's considered bad. My father never learned to speak Catalan because when he tried to speak it people made fun of him. But he considered Portbou to be his only home. [Do you know how to read and write Catalan?] I was educated during Franco and have trouble with reading and writing. I'm learning from my sons! [Is Portbou different from other villages in Catalunya?] I suppose it is because it's on the border. The family of my mother lives in the interior of the province of Girona and there the accent is very different. [If someone asks you your identity what is your response?] In Spain I'm a Catalana; in a foreign country I'm Spanish. I am a Catalana because I was born here. That's all. I'm a *Xarnega*. [What do you think of the new language law?] I think it's a bit rigid. [How do you see the future of Portbou?] I think it's very bleak. The population is dropping, there are too many retired people. There is little work in the village. There aren't enough hotels. There aren't enough activities either. Also, in the past many foreigners came here to shop because it was cheaper, but now that's over. Many people would like to stay here, but the housing situation is also very difficult. The road is a very real problem. I know someone who wants to live here and works in Figueres but has to begin work at 6 a.m. There are many old houses here that are closed up and in need of repair. Also, all the houses by the seaside are in bad shape because they are damaged by the salt air. [Are the festivals here important for the life of the village?] Yes, very. I think it's good for relations among the people of the village. It's the only thing that unites everyone, young, old, rich and poor, educated or not.

The Questionnaires

It is obvious that our questionnaires, like the interviews, do not constitute a random sample of Portbou's population. They do, however, add something to the ethnic and linguistic profile of the village particularly among younger people, of whom none were in the sample interviews that Sonia and I conducted. These questionnaires point to the common knowledge and use of Catalan in the younger population. In this respect it is worth noting that one of the two interviewers questioned mainly younger individuals. He himself is in his thirties and is fully bilingual. Additionally, he has many friends among those in Portbou who prefer to speak Castilian with members of their own social group, yet, during the interviews, most of those questioned chose to answer in Catalan. Also, the overwhelming number of these youngsters came from mixed Catalan-Castilian families. This is in fact a consistent pattern in the entire population of the village even among early arrivals. One obvious reason for this is the fact that many of the immigrants who settled in Portbou were single men who sought out Catalan wives from the local population. In the opposite case, where the mother was Castilian and the father Catalan, and, the children's primary language is declared to be Catalan, the children turn out to have spent much of their time with their Catalan-speaking grandmother. Eight of the interviewees identified themselves as Catalan and thirteen said they were Castilian. The rest gave mixed answers such as: "I am both Catalan and Spanish," or "I am Catalan and Castilian."

Chapter Five
A New Direction

At the end of our sabbatical (1996–1997) we had learned a great deal about the meaning of Catalan identity. We had also learned, to our dismay, that our planned comparative study was flawed by the reality of the situation on the ground. One sees only traces of Catalan culture in Cerbère. As for Portbou, while it is true that, in some ways, it remains a typical Catalan village, in other important ways it differs from the ideal because of its overly large transient population and the number of its resident retirees that severely skews the demographic pyramid. On the other hand, fortunately for Portbou, because it is in the region of the *Principat*, it can rely for strong financial support on the autonomous government in Barcelona that has a vested interest in maintaining Catalan culture in all the ways it can. As previously noted, the *Generalitat* has heavily subsidized the new port, the new school building, and the restoration of the Civic Center, and it has also expressed an interest in supporting projects associated with the Walter Benjamin Foundation. Beyond this local support, the central Spanish government has recognized the need to improve the economic isolation of Portbou by subsidizing the construction of an expensive tunnel between the village and Colera. This project, when finished, will make it easy for residents living in Portbou to drive to and from Figueres where work is relatively plentiful.

In the light of our original goals, Cerbère turned out to be even more of a disappointment than Portbou. Its cultural identity as Catalan, although strong in its early years, is now less than doubtful. The only major similarities between the two villages lie not in the cultural sphere, but rather in their sad demographic histories. Remember, Cerbère has, like Portbou, both an overly large retired population and a high proportion of transient workers, many of whom have no interest in putting down roots there. Even more than Portbou, where transit companies and their employees were never as important as in Cerbère, the latter has suffered greatly from the loss of these and other sources of employment that disappeared with the personnel reduction within the customs bureau, together with the significant diminution of the formerly large contingent of gendarmes. Most of the jobs for temporary residents are on the railroad but many of those so employed, because of the high cost of housing, live in other villages and commute to work in Cerbère.

Even on the symbolic level, the power of the highly centralized French government has weighed heavily on Cerbère. Although it has traditionally exerted negative pressure on local cultures and languages throughout the country, this pressure has been exceptionally pronounced in Cerbère. As Josep Pla pointed out (see my citation in chapter three), the French, while allowing Banyuls and other neighboring towns to maintain Catalan traditional architecture, promoted a distinctively Jacobin style on buildings in Cerbère to clearly mark the fact that it is French and not Catalan or Spanish. Moreover, Cerbère has, unlike Portbou, not been able to rely on the French government to subsidize the vestiges of Catalan culture that survive, at least to some degree, in its region.

In the current situation, cooperation between Portbou and Cerbère is inhibited by the small number of people on the French side of the border who are truly invested in Catalan culture, the continuing reticence of the French government to allow, or even encourage, cross-border economic or educational cooperation, and differences in law and custom between France and Spain that might appear trivial in other contexts. Finally, it appears that parents in Portbou are now more interested in having their children learn English than French since they believe that the former language has an advantage over French in their children's future. This is, perhaps, the major reason, but surely not the only one, why the trilingual school to be built on the former border between Spain and France, and developed jointly by the two villages, failed to see the light of day.

It is for these reasons that, to continue a comparative cross-border study of Catalan culture, we needed to extend our research into the past and current state of *Catalanitat* throughout the *Principat*, and to a lesser extent beyond that to all the Catalan-speaking regions of Spain, as well as to the French department of the Pyrénées Orientales. One positive result of our early research, however, was the fact that, in spite of their common special features, both Portbou and Cerbère gave us a preliminary insight into what today constitutes the contemporary Catalan experience under the umbrella of the autonomous region of the *Principat*, on the one hand, and French Catalunya, on the other.

Extending our study beyond the confines of two small villages, however, has been a daunting task. Anthropologists are used to working with relatively small, localized populations even when their studies are focused on national cultures. Our reorientation would have to be designed to collect enough data to reach what satisfied us as reasonable conclusions concerning the project. Parts of the data on which the rest of this book is based are the result of the need to develop a research strategy that would lead us well beyond traditional anthropological fieldwork. Looking back at what has transpired since 1997 we were lucky to have been aided by a combination of serendipity and planning.

The first necessity for our work was to build a substantial library focused on Catalan history, economics, culture, and linguistics. Very few Catalan books can be found even in university or public libraries in the New York area. We were obliged to build our own collection of printed source material. Our subscriptions to *Avui* and later to *El Temps* have been important in constructing an archive tied to contemporary social and political affairs in the *Principat*. But we were hard-pressed to find a good source of books. This is where serendipity came in. In 1997, I met a half-Catalan, half-American undergraduate at Columbia, Jordi Bricker, who was taking a course in our anthropology department. At the time, his father lived in the United States and his mother in Girona, a small Catalan city about thirty miles from the French border and on the rail line that connects Cerbère and Portbou as well as other villages and towns to Barcelona. Jordi agreed to introduce us to his mother and give us a tour of the city. We were to meet there the following summer on our way to Barcelona for a short stay. As it turns out Jordi was unable to meet us at the train station (he had an important appointment elsewhere) but we decided to spend the night in the city anyway. After finding a hotel we set out to explore and, in the old section of town, found a bookstore, Les Voltes, the only such *llibreria* in Girona, and perhaps throughout Catalunya, specializing exclusively in Catalan books on all subjects. For the following four years we continued to shop at Les Voltes buying contemporary books while browsing through the store's extensive collection as well as selections from those discussed in the literary sections of *Avui*.

While in Girona we also discovered to our pleasure that Catalan was the language most heard on the street, a marked contrast to Barcelona and other large cities through which we had passed on our visits to other parts of the *Principat*. Over the years since our chance visit to Girona, we have developed a strong friendship with the manager of Les Voltes and her husband. In 2001 we decided to spend three weeks in that city for intensive private lessons in Catalan. To this end we needed to find lodging with a family willing to share their house and their meals with us. This is not an easy task anywhere in Catalunya where most people are reticent to share their lives with strangers or even friends. The house is the center of intimate family activities and what entertaining takes place is usually with other relatives. Although we have warm relations with many individuals and families in Portbou, only a few have ever invited us to their homes. In questioning informants we were frequently told that inviting nonfamily members to one's house is a rare practice indeed. We were honored when this rule was broken for us as it was in Portbou on three or four occasions and, in Girona, by Rosa, the manager of Les Voltes and her husband, Esteve, an architect who works for the suburb of Salt just outside of the city. It has become routine on our regular visits there for a few days every fall to alternate regularly between dinner at their home and eating out with them in one of our mutually favorite restaurants.

In the fall of 2003 on one of our regular visits to Girona, Rosa and Esteve informed us of a crisis that had occurred that summer concerning the bookstore. The building in which Les Voltes is located is owned by the Catholic Church. During the summer of 2003 the bishop of the Girona diocese decided to consolidate the several offices of the bishopric and put the clerical bookstore in the space occupied by Les Voltes. The only alternative to expulsion was an untenable rent increase, which, if paid, would quickly bankrupt the store. Furthermore, there was no other location in the commercial part of the city available at a reasonable rent. (The store, in fact, generally ran a deficit that was covered by an association created by its founder, Feliu Matamala, in 1963.)

Although certain Catalan books had been allowed under the strict censorship of the Franco regime, Matamala did what was, at the time, an unheard of, daring act just after the opening of Les Voltes. He challenged the government by putting Catalan books on display in the store window. In subsequent years Les Voltes became a center of Catalan resistance to Franco and, later when democracy was restored to Spain, provided strong support to actions designed to counter any acts, private or governmental, that could be considered hostile to Catalan culture. For example, in the beginning of this century, the central government made it obligatory, in accordance with European Community rules, for all automobile license plates to bear an E for Spain [España] on a blue background. Les Voltes immediately joined a movement to cover the E with a bumper sticker bearing a CAT. Anyone wishing to put such a sticker on his car could have one free of charge at the bookstore. At the same time the ERC (Esquerra Republicana de Catalunya) offered to pay the fines of any person fined for making this change. Les Voltes has also become the producer and distribution point in Girona for obtaining unofficial Catalan identity cards, a nonviolent demonstration against the hegemony of the Spanish state, which costs more to produce than the modest fee the store asks for them. On a table next to the bookstore entrance one can also find printed information, often accompanied by a petition requesting signatures, concerning current Catalinist activities in Girona.

When news of the immanent closing hit the newspapers, the local Catalanist community responded with an enormous and spontaneous protest demanding that the bishop relent. The sale of identity cards tripled in a just a few days and the press began to carry stories about the dispossession, as well as angry letters from readers against the action. The protest gathered steam throughout the summer months but the bishop would not relent. In the early fall Mr. Matamala was interviewed on a local TV station. We saw a taperecording of this program and were impressed by his passionate and articulate protest against the elimination of the bookstore. Finally, in the late fall, one week before the expulsion was to occur, the bishop relented and agreed that as long as Mr. Matamala continued as the director, the store could remain open in its original place. Though this might seem generous to some readers, I must hasten to add that, at the time, Mr. Matamala was ninety-one years old. Yes, fortunately he is, as of this writing, in good health. For example, at the beginning of the fall in 2003 he drove his car alone all the way to Perpignan for the opening of the first center for Catalan culture allowed in the department since Northern Catalunya had been incorporated into the French state.

Girona, where Catalan is very much alive, was indeed an ideal place to continue to improve our language skills. But how to find a family with which to live as well as share meals? Locating a language school that would offer us private lessons several hours a day would be much less of a problem. We appealed to Rosa for help in finding compatible housing. Well connected to the Catalanist community in the city, she generously offered to aid us in our search. This effort, through no fault of her own, was unsuccessful even though she spent a good deal of time trying to fit our needs. As a last resort we turned to another Catalan connection we had in Girona. This was the woman who had helped us before in our search for lodging in Barcelona during our first visit there in the winter of 1997. She suggested a family living just to the east of the city that, as a favor to her, was willing to provide us with room and board for a reasonable fee. They are a very interesting couple. He is Catalan and his wife, who speaks fluent Catalan, and generally speaks it in the household, is Spanish from the region of Extremadura. They are both of working-class origin and are employed in the post office of their town. Their household, with its four children now grown and living elsewhere, is Catalan-speaking although they, like the vast majority of Catalans, are bilingual. In the fall of 2001 we joined them, participating in the family's evening meals and doing small chores around the house. They too have become fast friends with whom we now share meals during our short stays in Girona, alternating between their home and the restaurant in our hotel.

Though these contacts have been important for us they did not solve our major problem. How could we, in a reasonable amount of time, get to know and speak with a large number of Catalans from the *Principat* as well as other Catalan-speaking regions? The solution lay in attendance at the Catalan Summer University held every year during the second half of August for ten days and offering a wide range of courses in Catalan language and culture. The *Universitat Catalana d'Estiu* (UCE) takes place in Prades (population 6,000) the French Catalan town in which the renowned cellist, Pau (Pablo) Casals, spent his exile during the Franco years and where Pompeu Fabra, the man most responsible for standardizing modern Catalan, died in 1948 and is buried. It was in Prades, too, that Casals founded the world famous music festival that bears his name, and that has taken place every summer since the early 1950s. Prades, by its location and association with Casals, who throughout his life was a strong Catalan nationalist, was the logical place outside of Spain where Franco still ruled to encourage Catalans from all Catalan-speaking areas of Spain and elsewhere to come together to learn about their history, literature, economy, and culture as well as to socialize and exchange ideas. The original inspiration for the school came from a group of young idealistic French Catalans who were influenced by the left-wing student uprisings of 1968. This stillborn rebellion, inspired by a flush of new radical thinking and action, also became a rallying cry for a renaissance of regionalism against the centralized power of the French state.

In its first year the UCE took place for only five days. It was limited to a single section and had 50 students. In 1978, ten years later, there were 300 inscribed with 22 sections offered. Ten years after that in 1988, there were 700 students for 24 formal classes, 9 seminars, 19 workshops, and 9 daylong topical sessions. By 1990 attendance had reached one thousand and, during our first participation in 2001, about fourteen hundred students attended. In 2003 inscriptions passed the fifteen hundred mark.

The UCE offers short but intensive courses taught by well-known professors from various Catalan universities, experts in Catalan language and culture from foreign universities, and professionals in a range of jobs in government and the private sector. Initially, we were hesitant to attend, thinking that our Catalan language skills were not yet strong enough to get much out of lectures and discussions and, I must admit, out of fear of embarrassing ourselves. By 2001, although still somewhat reticent, we decided to try our luck. To our surprise, beyond the fact that the UCE turned out to be a rich source for learning about Catalunya, it also provided an unusual opportunity to meet and speak both formally and, more often, informally with a wide range of people of all ages and professions who identified strongly with Catalan culture.

The UCE takes place on the campus of the local *lycée* (high school) perched on a hill overlooking Prades fifteen minutes on foot from the hotel in the town center, in which we stay. The *lycée* faces the impressive Canigó Mountain rising to the south above the town to a height of about 10,000 feet. As already noted in chapter one, this mountain is a major symbol of Catalan unity, providing a dramatic backdrop for the activities of the university. This symbolism is celebrated by a hardy group of students and professors every summer toward the end of the session in a long climb up the mountain where they pass the night and, in the morning, celebrate sunrise.

Although expert professionals rigorously teach the courses, the atmosphere outside of the classroom is quite informal and relaxed. Practically everyone, teachers and students alike, eat together in the *lycée* cafeteria. We soon became used to sitting down at lunch and dinner, each time with a different group of students and/or professors to engage in serious conversation about Catalunya and the contents of the various courses we attended.

Classes are organized in the catalogue around traditional and nontraditional university departments. The following major subject categories we found in the 2001 course list and, in general, represent the wide choice of material followed each year; *Pluridisciplinary Studies*—thirteen courses; *Natural Sciences*—eleven courses; *Science and Technology*—three courses; *Social Sciences*—six courses; *Economy*—eight courses; *Politics and law*—three courses; *Communication*—four courses; *Humanities*—nine courses: *Language*—seven courses; *History*—five courses; and *Art History*—five courses.

Among the courses particularly useful for this study—of the many we attended—were two separate offerings by the sociolinguist, Albert Branchadell. One, taught in 2001, traced the European Community's language law concerning the rights of minority languages spoken within national borders, which either have no legal status under existing local law or were recognized legally, before the European law was passed by its parliament in those countries that have also recognized multinationality within their borders (Switzerland, Belgium, Finland, for example). During the following summer course (2002) Branchadell outlined for the class the varied reception of this law received in different member states. While most agreed to accept all its provisions, both France and Spain only gave it partial acceptance (see chapter six). Branchadell's course presented an analysis of the status of Catalan in different parts of Spain, as well as Andorra and France. Another course taught by a high-school teacher in the department of the Pyrénées Orientales, Alà Baylac i Ferrer, dealt with the social and political history of Catalan culture in the region of *Languedoc-Roussillon*. The German philologist, Tilbert Stegmann, who is interested in improving linguistic competence in foreign languages within Europe, offered a method for *reading* all of the Romance languages in six lessons! As bizarre as this might sound, the method is highly successful and, although further work is required outside of class, provides a useful introduction to the subject. Alan Yates, professor of Catalan language and literature at Sheffield University in England, and author of the first book for English speakers for the self-teaching of Catalan, offered a course on the poetry of Jacint Verdaguer, the most important Catalan poet of the late nineteenth century. While these courses inspire special mention, we found most of the other classes we attended not only useful for this study but also for our general knowledge of Catalan culture.

Courses vary in length from one to three hours per day and from one to six days, although a very few take up the entire ten days. Thus, in general, it is possible to take several subjects during the summer's offerings. Although the major departmental categories listed above tend to repeat from year to year, as they would in any university, except for some particularly popular subjects taught by well-known professors, specific courses are not repeated until several years have passed. Thus, attendance at more than one summer session allows repeating students to accumulate knowledge in several subjects. Individuals officially enrolled in Catalan universities can receive credit from their various institutions for enrolling in the UCE.

Beyond its wide-ranging set of courses the UCE offers a series of other formal and informal activities. Between noon and half-past one every day there is a plenary session dealing with a single topic. These range from speeches by experts involved in current affairs of interest to Catalans, to sessions honoring social, political, and cultural (most often literary) figures. In 2003 these assemblies were devoted to the coming Catalan parliamentary elections. The desire among Catalans of all political parties to strengthen the existing statute of autonomy for the *Principat* was emphasized. The participants present at these sessions spoke for the PSC (Catalan Socialist Party), the CiU (*Convergència i Unió*, the center-right

Catalanist party), the ERC (*Esquerra Republicana*, the left independence party), and the IC-V-E-U (*Iniciativa per Catalunya Verds-Esquerra Unida*, the former local Catalan communist party in alliance with the Green Party and the former Spanish national communist party). The PPC (*Partit Popular de Catalunya*, identical to the right-wing Spanish nationalist party, the PP) was not represented because it opposes radicalization of the existing governing statute for Catalunya.

The university also offers daylong lecture-discussion sessions under the rubric of "Science and Professionalism." These range over a wide variety of subjects and allow for intensive formal and informal treatment of specific topics. In 2001 these included Andorra and its relations with its neighbors to the south, as well as the problem of environmental pollution.

Lunchtime is usually between twelve-thirty and two in the afternoon, a compromise between the French and Catalan eating hours. The meal is followed by a series of workshops including such projects as art, learning the Sardana, hiking, choral work, theater, photography, medicinal plants, speech improvisation, tai chi, circus performing, comic drawing, and special sessions for young children of adults inscribed in the program.

Between six and seven-thirty every evening there is often a second plenary session dealing with literary, scientific, cultural, and political issues. Additionally, on most afternoons, a wide range of more informal presentations and discussions occur dealing with recently published books, current political activities, and other matters pertaining to Catalan culture throughout the Catalan-speaking world. Dinner starts at seven-thirty (very early for Catalans but food is still served until at least eight-thirty) and is followed by a two-hour concert in the central public square of Prades. The entertainment includes folk, rock, and popular music featuring performers from various regions where Catalan is spoken. Basque musicians are often also invited to perform. If this is not enough to fill the daily schedule, every night a Catalan film is presented in the Prades Cinema. We, however, have never attended any of these sessions since at our age we tend to fall into bed about the time the film begins. Finally, it is worth noting that fees for the university are kept very low. Every year the sessions receive financial contributions from a number of sponsors, including the Generalitat of Catalunya, the government of Andorra, the Region of Barcelona, the City of Tarragona, the Governing Council of the Pyrénées Orientales, the Governing Council of the Island of Menorca, the City of Barcelona's Institute of Culture, the Regional Government of the Balearic Islands, the Regional Council of Languedoc-Roussillon, the Region of Lleida, the Region of Girona, the Council of Mallorca, the Island Council of Eivissa (Catalan spelling) and Formentera and La Caixa (the major savings bank of Catalunya, and an important donor to cultural activities). Low cost housing and food are provided in the *lycée* dormitories and cafeteria. Four individuals, however, must share rooms available in the dormitories,—a fact that leads us to take our residence in a local hotel.

Given the number of young people who attend the UCE, of the two on-campus dormitories, one is reserved for young participants and another, more isolated from the main area of the school, is reserved for older people who wish to enjoy a good night's sleep. There is a reasonable package price for tuition, plus lodging and meals at the *lycée*. For tuition alone, cost was only one hundred fifty euros per person in 2003. Numerous scholarships are also available for high school and university students from all the Catalan-speaking areas including French Catalunya and the city of L'Alguer. Since the summer of 2001 we have attended each yearly session. Where else can one meet and interact with 1,600 *Catalanistes* from the entire Catalan-speaking world who represent a wide sample of ages, professions, and social backgrounds?

In 1993, a convention was signed formalizing the existence of the UCE and its place for dialogue between Catalunya and Llenguadoc-Rosselló (Languedoc-Roussillon), the region composed of a small set of departments in the south with its own elected government. Although these new bodies represent a limited form of decentralization in France, they remain essentially under the control of Paris. On the other hand, under the new system, local culture does have a more important, if limited, role to play in the life of regional France. The signers to this agreement were Jordi Pujol, president of the *Generalitat*, Jacques Blanc, president of the French regional government of Languedoc-Roussillon, Paul Blanc, senator in the French parliament and mayor of Prades, and the UCE represented by its president, Max Cahner, as well as the president of the Foundation for the UCE, Emili Giralt. What follows are extracts from the *Convenció* (Convention):

> Prades has the vocation to become a center for cooperation and dialogue between Catalunya and Languedoc-Roussillon in accord with the principles of the European Community concerning trans-frontier areas. Since its creation in 1968 the UCE has played a role in the promotion of the culture, language, history, and the development of science in all Catalan-speaking areas [The *Països Catalans*]. It participates in interchanges between the world of Catalan universities and the intellectual and scientific domains in the rest of Europe. (My translation from the Catalan text)

In recent years there have been some conflicts, not only among academics in Catalunya, but also in other Catalan-speaking regions, over the status of the UCE as a center for the expression of Catalan militantism in its many forms versus its existence as a purely academic institution. So far, the tradition of Catalanism has been maintained. It needs to be stressed, however, that this has occurred in the context of a course structure that, if not apolitical, meets the highest standards of academic freedom, including openness toward conflicting viewpoints. The representation of the Catalan world in all its richness has thus included not only its language and culture but also an active militancy in favor of a range of options concerning desired changes in the political structure of Catalunya. These vary from a desire for greater autonomy to complete independence for Catalunya. As an example of an intermediate position it seems worth noting the point of view of the sociolinguist Albert Branchadell regarding linguistic matters in the *Principat*. Although he is engaged in the struggle for the preservation of Catalan as a one of the four official languages of Spain, and works actively for establishment of the Catalan language on all Spanish passports, drivers' licenses, stamps, and other official documents issued by the Spanish government, he opposes independence for Catalunya, as well as laws intended to exclude Castilian from its co-official status in the *Principat*. He fears that independence could have negative effects if Castilian speakers were to make *special* demands to "preserve" *their* language under the European Community's minority language law.

The reception of the UCE in Prades runs from quite friendly (it offers a good deal of additional business for cafés and restaurants) to indifference but there seems to be little active opposition to it. Many local residents happily attend the musical events on the main square. Every year the mayors of Perpignan and Prades attend the first plenary session to offer a warm welcome to the assembled crowd. Sometimes other elected personalities from the region also speak. But not all reactions to the school are positive, at least within Spain, and even Catalunya. In the summer of 2003, two articles, one copied from the far right-wing Spanish newspaper, *La Razón*, and the other from the hispanophile and monarchic Barcelona paper, *La Vanguardia*, were posted along with other, more positive, newspaper reports concerning the activities of the UCE. What follows is extracted from each of these two papers. In *La Razón*, under three photos (one of Pasqual Maragall, the head of the PSC (Socialist Party of Catalunya), another of Ibarretxe, the Basque (nonviolent) nationalist leader of the current Basque government) and the last one a strange shot of Jacques Chirac, the president of France standing behind a bank of microphones with a wide grimace across his face, expressing his disgust at *something* of which no one can be sure. The text goes on to call both Ibarretxe and Maragall "Spanish" nationalists (an interesting fact because the central government in Madrid is, in general, allergic to the term "nationalist" as a self-referent). Apparently the paper dares not use the terms "Basque nationalist" or "Catalan nationalist" (which Maragall is not in any case) and so through a peculiar use of self-censorship substitutes "Spanish" for Catalan and Basque. All this was in response to Maragall's call during a plenary session of the UCE dealing with plans for a new statute for Catalunya. Maragall, far from pleading for independence, instead called for a cross-border economic axis among the Catalan Mediterranean autonomies of the *Principat*, València, and the Balearic Islands, along with the region of the Midi-Pyrénées in France and Aragó. Ibarretxe was not even present at the UCE and, as far as I know, has never suggested that Basque independence, which he clearly favors, will include the Basque area of France.

The article in *La Vanguardia* by Manuel Trallero concentrates on painting a rather dismal picture of the UCE's physical appearance noting that "while in the town there are posters advertising the event, there are no instructions concerning how to arrive at the location of the University as if it were a clandestine and illegal event intended only for the initiated . . . it is easier to find the location of a tennis match, or an antiques fair. . . . When one finally does find the University, the disillusionment is total. Again there is no sign posted to indicate that one has arrived. It is like a reunion of Christians in the catacombs of Rome in the time of Nero with the lions circling around them." Though directions to the university could have been a bit clearer, it's obvious that the writer is taking advantage of this minor oversight to clothe the entire event in heavy parody. Even the few courses and other activities of the UCE that are mentioned in this portrait are seen only in a negative light.

In 2001, a young professor in the Spanish and Portuguese department at Columbia University, born in the United States of Catalan parents and trilingual in English, Castilian, and Catalan, began to offer courses in Catalan. We were allowed to sit in and also attend an extra session each week during the academic year for Catalan conversation. We have continued with the weekly sessions ever since, which give us the opportunity to practice the language with a range of people, including those who are native speakers and others who are quite fluent in the language. Another opportunity to touch base with Catalans in New York is through *Els Amics de Catalunya* (Friends of Catalunya) a group that meets every Friday evening during the year at a midtown hotel managed by a person of Catalan origin. The hotel owner is also a Catalanophile. *Els Amics* is essentially a social club for young professional Catalans who work either temporarily or permanently in the New York area. Some are students at various universities in the city and elsewhere. We have become quite friendly with a few of them and, along with other Catalan-speaking acquaintances, we invite them to dinner at our apartment. Most of the individuals with whom we interact in New York are Catalanists and identify strongly with their homeland. Discussions around the dinner table usually involve the current political, social, and economic situation in Spain and the *Principat*.

In the summer of 2000, we had the good luck to meet a young Catalan couple while working as volunteers in La Jasse, the cultural and commercial center of the Larzac community that we had previously studied (see chapter one). Sonia overheard the couple speaking to each another in Catalan and immediately asked (in Catalan) where they were from and what they were doing in France. It turned out that they are from Balsareny, a town just below the foothills of the Pyrénées, in the back country of Catalunya. They had just come from Brittany where they had seen the total eclipse of the sun that was visible there that summer. They were surprised to find Catalan speakers on the Larzac, particularly two Americans, and we exchanged addresses. It turned out that they frequently camped out in France during their summer vacations since their financial means were very limited. In fact, while they spent their summers as a couple, at the time, because only one of them was employed, they lived separately during the year in the houses of their respective parents. The following year, again camping in France, they dropped in at our house at the foot of the Larzac. I was out when they arrived, and by the time I returned from shopping they were gone. It was a lucky break for us that Sonia was home to receive them. On my return we discussed their visit and quickly decided to invite them the following summer to share our house and food in exchange for conversations in Catalan. They accepted and, for the next two summers, we discovered them to be excellent house guests, sharing chores and their experiences as working-class Catalans with us. Finally in the fall of 2002 they were both employed, had found an apartment, and reciprocated, as they had promised, with an invitation for us to visit them in Balsareny. Gracious hosts, they took us to visit the nearby cities of Manresa and Berga both of which, as inland communities, have high percentages of Catalan speakers. Berga is also the home of a spectacular summer festival, the *Patum* described as follows by Dorothy Noyes:

> Performed annually for nearly four hundred years the Patum of Berga has simultaneously celebrated and refused the political order at every turn.

It's dancing effigies—giants and dwarfs, Turks and Christian knights, devils and angels, a crowned eagle and two flaming mule dragons—serve as vehicles for a multitude of allegories. But the festival obscures its own apparent messages through techniques of the body cultivated with special intensity since the last years of the Franco regime: strong rhythms and constant motion, vertigo, heavy drinking, sleep deprivation, and the smoke and dense falling sparks of firecrackers at close range. After five days, the symbolic combat ends in physical consensus and the incorporation of both individuals and social categories into a felt totality: metaphorically a social body. (Noyes, *Fire in the place*, 1)

Unfortunately for us (or perhaps fortunately!) we were not present for this (perhaps most famous) Catalan festival. But we did see the miniscule space, a mini-*plaça*, in which the actions take place and could only imagine the packed in crowd, the fire, the noise, and the weary drunks who have from the beginning imbibed a special mixture of alcohol guaranteed to inebriate the drinkers immediately and lead at the end to a major hangover.

Though we would have been happy to visit these two cities in the heart-land of Catalunya at greater length, even without the historical presence of the *Patum*, the most interesting excursion for us was to a former company town, not far from Balsareny, built to house and provide for the needs of its employees. This was one of the very few textile plants that still functioned in this now depressed area at the time of our visit, even if with a much reduced workforce. In former days these towns were the closest things you could get to a coercive "total" institution. Controlled by capitalist barons, who ruled almost every aspect of their workers' lives, they served to fuel the economic miracle that was industrial Catalunya in the nineteenth and early twentieth centuries. Most factories are now closed, victims of the flood of "outsourcing" that has affected Catalunya's traditional industries just as it has other formerly strong economies. Even this factory closed not long ago, but was saved as a community by new owners who then sold off all the nonindustrial spaces on the property and reopened a small part of the plant with a much reduced work force. The housing stock was purchased as a block by the governing body of a nearby town, which then sold the apartments to individuals, many of whom were retired workers from the plant. Some were bought as secondary residences.

Of the couple who were our hosts in Balsareny, the woman introduced us to her father, a former worker in the plant who had purchased his apartment shortly after retirement. She had grown up there, sharing a small two-bedroom flat with her parents and eight siblings! Her father and one of his friends graciously took us on an extended tour of the entire community. Under the traditional organization of the plant, in addition to the actual factory, the complex was equipped with a store that sold items of everyday necessity for the exclusive use of residents, a café, a restaurant, a library stocked with religious tracts, a church, and a fully equipped theater. Some of the communities, if not the one I am describing, went so far as to print their own script that served as money within the factory compound so that in general the workers did not need to have national currency. During our tour we learned that the owner of the entire complex lived off the grounds in a large mansion and delegated the direction of the facility to a paid manager, who was not a relative and who lived among the workers so as to supervise them better.

The theater and the church were both parts of a vast propaganda machine designed to control the workers. Attendance at the mass was a requirement of employment. Slackers risked losing their jobs if they were absent, except for illness, from the weekly services. The priest, in addition to his position as a moral authority, was an important political figure who followed the most conservative teachings of his regional bishop. The theater served as the setting for propaganda against unions and radical political parties. Plays were limited to religious themes representing the eternal struggle between Christianity and the Devil. The "evil ones," figuring in the *Patum* celebrated nearby every summer in Berga, and familiar to the members of the community, wore color-coded costumes: red, the color associated with the communists and socialists, or black, the color favored by anarchists.

Although the main focus of our study is Catalan culture in the *Principat* and Roussillon, we did additionally take advantage of a week's summer school in Maó (Catalan spelling), the capital of the island of Menorca (Catalan spelling) one of the four main Balearic Islands. The summer program in Menorca, bearing the imprimatur of the distinguished University of Mallorca in Palma, the capital of the largest island in the Balearics, offered a few courses that seemed useful for expanding our general knowledge of Catalan culture beyond the confines of our principal focus. Furthermore, the island was less daunting than elsewhere in the Balearics, given its size and rather homogeneous population in comparison to the other three main islands that comprise the region. It also appeared to be a better place to gather useful information on the status of Catalan culture in this autonomous region than the large, crowded, and touristed Mallorca, which also has a sizable and permanent population of ex-patriot Germans.

There are two other main islands in the Balearics, smaller than Menorca—Eivissa and Formentera—but the latter is miniscule with the smallest population in the archipelago and the other is overpopulated by tourists and discotheques. As citizens of the European Community, Germans, as well as people from the other countries of the European Community, have the right to live and work in any country that is a member. Palma has at least one German newspaper as well as several German-owned stores and banks. It is a sad joke among some Mallorcan Catalanists (whom we met at the UCE) that the most frequently heard language in Palma is German rather than either Catalan or Castilian. Thus, because Menorca was best suited for our needs and as it offered a useful structure for meeting Catalan speakers, we found ourselves once again students enrolled in a weeklong schedule of courses, taking both the morning and afternoon programs.

The summer session takes place in the upper floor of a former monastery, a very pleasant location at the edge of town, perched on the low cliff that overlooks the vast bay below. The classrooms are located on two sides of, and look down on, a former cloister. The ground floor of the building has been converted into a shopping center and the basement, under the former cloister space, has become a large supermarket. The formal name of the school is *Universitat Internacional de Menorca Illa del Rei*. This is a somewhat pretentious appellation since the only activity it has at present is the one-week academic session in the month of September. It has no full-time faculty and its teachers, all distinguished in their fields, are drawn from other universities in the Balearics, the *Principat*, and the rest of Spain, and sometimes from foreign countries.

As it turned out, though our stay in Maó was fruitful, there were certain shortcomings for us in the way the sessions were organized, particularly in comparison with the UCE. First of all, students do not eat meals together and lodgings are scattered in different private accommodations. While there are morning and afternoon breaks that last about thirty minutes, there is not much time or space to socialize and get to know the participants. Equally important is the fact that although the brochure outlining the formal offerings for the week was exclusively in Catalan and many of the courses to be offered appeared to fit our needs, it turned out that quite a few of them were taught in Castilian, a language I have not mastered enough to understand in the context of a crowded classroom. I even have difficulty reading it. (In fairness to the school, I should mention that the Online Catalogue of the UIMIR does say that the official languages of the school are, in addition to Catalan, Castilian, French, and English, but it does not make clear which courses will be taught by non-Catalan speakers—many of whom during the summer of 2003 taught in Castilian.) This impediment immediately cut down on what sessions I could attend, although Sonia, more familiar with Castilian, was able to sit in on a few subjects of interest to her. I, for example, was interested in taking the following offerings that were enticing but that turned out to be closed to me because of my linguistic incompetence: The Social Imaginary, the Cultural Construction of Reality; the Notion of the Island in the Context of Literature and Multiculturalism; and Nationalism in the Context of Modern Democracies. On the other hand, a very useful course taught in Catalan and which we both took was the Demography of Mallorca that dealt not only with the current population of that island but also with the social problems associated with rapid immigration, as well as the problems concerning social services in relation to a increasingly growing older population.

Although we had almost no time to spend outside of class, which took place between nine in the morning and two in the afternoon and then again from four in the afternoon until seven in the evening, we made good use of what ever free time was available to explore Maó, both in the lower touristed bayfront and in the town above. The surprising thing about Maó, a small pleasant city with a population of only 20,000 (the island is largely agricultural with a large dairy industry), is the degree to which one hears Catalan (or Menorcan, as the natives prefer to call it) spoken everywhere except in the port full of tourists who dock along the bay and who tend to either stay on their boats or visit the restaurants that line the area for about two miles. They do also venture into the upper town, but generally are seen only in the busy business section in the city center, either in shops or in the many cafés. Not only is Catalan the language commonly used, but it is also used by people of all ages, unlike Barcelona where we frequently heard Castilian spoken by groups of children and teenagers who appeared, in spite of their schooling, to prefer Castilian to Catalan. On the other hand, while the Barcelona Catalan-language daily newspaper, *Avui*, is available every morning on the news stands in Maó, the two local Menorcan papers publish primarily in Castilian. One of them has a two- to four-page section in Catalan as well. Although Eugeni Casanova, in his book, *Viatge a les Entranyes de la Llengua: De la Riquesa a L'agonia del Català* (Voyage into the Inner Reaches of the Language: From its Riches to its Agony) claims that the only Catalan books he found in book stores in Maó were about tourism on the island, I did find two that had a fair selection of Catalan books along with a larger number in Castilian. On our way to the airport to leave the island for Barcelona, I spotted a bookstore somewhat eccentric to the center of Maó that bore a sign indicating that it sold Catalan books exclusively. However, we have no idea of the size of their stock or the subject matter offered. Yet it does appear that there are Catalan readers, at least to a moderate degree, and in spite of the lack of a local Catalan newspaper, the spoken language seems to be thriving. On the other hand, there does not seem to be much faith in the Catalan language leading to commercial success. I looked over the books for sale in the airport bookstore and only found one Catalan book. It was a translation from English of Michael Moore's *Stupid White Men*!

Chapter Six

The State of Catalan in Rosselló (Roussillon)

How Northern Catalunya Became a Part of France

In the year 1659, Louis the Fourteenth of France signed the Treaty of the Pyrénées with Spain. The Catalan Parliament (*Les Corts Catalanes*) was not a party to the agreement. In principle, the treaty was created to establish a "natural" border between the two countries defined by the Pyrénées mountains, a fact belied by a division drawn across the center of the great plain of the Cerdanya plateau, putting half of it under French rule while the other half remained under the titular rule of Spain. At the time of the signing, and for many years thereafter, the region on both sides of the border remained culturally and linguistically Catalan. After the French Revolution when France was divided into departments, this large Catalan territory, with its capital in Perpignan, was folded into a new administrative division, the Pyrénées Orientales, which, in its northern portion, included a region whose inhabitants spoke Occitan. This was part of a general pattern in France, which, in creating these political divisions, never allowed traditional linguistic areas to fully match the new political divisions. This is as true of the Basque country and Brittany as it is of the Pyrénées Orientales, and remains today a bone of contention between Paris and those regions that might want to create culturally homogeneous political units. Still, in the initial stages of French governmental administration in the Pyrénées Orientales, little was done to suppress the Catalan language. It was, in fact, widely used until modern times when, at the end of the nineteenth century, French education was secularized and placed under the control of the central government. It was in the newly created public schools that speaking Catalan was formerly prohibited, but even then it maintained a certain degree of dominance outside of the schools, in everyday social relations, until the 1960s. As we shall see a little later, it was a great tide of out-migration by natives of the region seeking work elsewhere under the pressure of the French bureaucracy as well as a recent tide of in-migration by retired people who found the mild climate and charming landscape a pleasant area in which to pass their declining years that undermined the use of Catalan in the region. By the time we began our research project we rarely heard Catalan spoken in public as we traveled around the Department.

Our Initial Search for Catalan Culture in the Department

In order to familiarize ourselves with the landscape of Roussillon and its culture we took short trips to the area in the first year of this new century and again in the following year. We have returned since to attend the UCE in Prades and have maintained contact, if less intensively than before, with the people of Cerbère. Only vaguely aware, at the time, that the UCE took place in Prades but fully recognizing that Prades was the chosen place of exile for the famed Pau Casals, we chose that town for our first visit. Although our stay there was short, our impression was that it was totally French in character. True the Catalan flag hung from the City Hall, but this is not unusual anywhere in Roussillon, and carries little significance beyond its attraction for tourists and has little reference to either Catalan identity or language use. Everywhere we went the commercial signs were in French, and French was the only language we heard in the streets and in the cafés we frequented. Of the two bookstores, one had a few books in the Catalan language, primarily concerning regional history, but the majority of the stock was in French. In neither of the two stores did the clerks speak Catalan.

During our several visits to the region we were told that, with the exception of Perpignan where, of all people, the Gypsy community is primarily monolingual in Catalan, to hear the language spoken we would have to go well into the southwestern reaches of the Department. For this reason we made plans to stay for a few days in Prats De Molló, which is in that corner of the region, and take day trips from there to neighboring villages and towns. We made a reservation in a hotel and on arrival discovered that the lady owner of the establishment and her elderly mother were indeed Catalan speakers. As it turned out, the younger woman represented a typical case of out-migration from the region in order to advance her career. She was accepted in one of the *Grandes Écoles* (elite universities in France where people prepare for professional teaching or business careers in science, the humanities, and industry). After graduating, she remained in the capital and did very well in her chosen business finally to return to Prats De Molló later in life to aid her aging mother in the daily affairs of the family hotel. She readily told us that it was the French language that had enabled her to succeed and that she was grateful for the opportunity the language had given her to integrate successfully into French culture. This was our first taste of an old story repeated over and over again by casual informants and interviewees.

From Prats De Molló, where again we heard very little Catalan spoken in public places, we took day trips to several small villages in the backcountry where we did find some Catalan speakers—generally, older people. Those few with whom we had brief conversations were eager to inform us of their love of the Catalan language, which they felt was dying in the region, at the same time assuring us of their total loyalty to France. Our last and, for us, most interesting visit on this particular trip was to St. Laurent de Cerdans not far from the border with the *Principat*. While there we first visited one of the only two remaining factories producing the typical Catalan cloth and rope shoes known in Catalan as *Espardenyes*. Here the few workers in this small enterprise all spoke Catalan and were amazed and delighted to speak it with two Americans. They explained that they were just managing to survive in the business due to the increasing competition from the sports sneaker industry, the products of which were so popular with young people. They told us that the town had two other attractions that might interest us. One was the only museum in France dedicated to the history of the refugees from the Spanish Civil War who, in the dead of winter in 1939, had crossed over the border in the Pyrénées to seek refuge in France. This museum was complete with reconstructions of what the barracks were like in the camps and preserved many examples of press reports concerning the migration of over 100,000 people, some soldiers and others civilians. The conditions in the camp near St. Laurent were primitive, however it did provide adequate if crowded shelter to the internees, but internees is what they were. I must add that other supporters of the Republic, including many Catalans, fled Franco by passing into France along the Mediterranean coast. These refugees did not fair as well as those who managed to get to Prats De Molló and St. Laurent. They were forced to live under terrible conditions, many of them sleeping out of doors in the sand of the beach at Argelès, now a tourist site and the gateway to the *Côte Vermeille*. Although in every sense of the term a concentration camp in both accommodation and treatment of the internees, the former minister of the interior under President François Mitterand, Jean-Pierre Chevènement, in commenting on a project to erect a monument to the refugees in Argelès, refused to allow the term "concentration camp" to be used on the monument because of its resonance with the Nazi camps of the Second World War. His statement was: "Concentration camps never existed in France." This statement is taken as truth by many contemporary French people who, to this day, refuse to accept the fact that most of the refugees from the Spanish Civil War were shabbily treated by France in 1939. Furthermore, many in France, even leading politicians, deny that the French-run Vichy government during the German occupation was in any sense a fact of French history. It was not until Jacques Chirac was elected president of France that a head of state declared that this aberrant view of history was wrong and that the Vichy regime was a bona fide French government and as such played a role in French history, and not a glorious role at that.

Our last short visit in St. Laurent was to a cloth factory producing tablecloths, napkins, and towels with typical Catalan designs. There were few workers in the now mechanized establishment. It was spooky for us to see the looms creating highly decorated items without the intervention of the human hand. In the store attached to the factory we found two young people. When we struck up a conversation with them in Catalan, they responded with a garbled mixture of French and Catalan. One or two other youths who were also present apparently found it very amusing to hear Catalan spoken. These, and other admittedly anecdotal experiences in the backcountry of Northern Catalunya, did not give us much confidence in the idea that the Catalan language was alive and well in the region. It turned out, as readers will see in the next chapter, we were partly right and partly wrong about what is actually a rather complicated situation.

A final trip to familiarize ourselves with the state of the Catalan language in the Pyrénées Orientales was to Céret, a town in the southwest of the Department, only a few miles from the border with the *Principat*. Our main purpose there was to interview a Catalan speaker familiar with the status of the language in her subregion. We spent two days in Céret and listened for Catalan in the streets, cafés, and stores in this picturesque town of about 7,200 habitants. No one in the venues we explored appeared to converse in the language either with one another or with customers.

When we inquired about the language, only a few people responded that they spoke it. Later we were told that we would be sure to hear Catalan in the once-a-week outdoor market where local food products, as well as cheap clothing and tools, are sold by itinerant merchants who follow the regular weekly cycle of markets held in neighboring towns and villages. Such markets are still common throughout rural France. It is a general rule that if such markets exist in an area with a regional language, even one deep in the process of disappearing, that will be *the* place to hear it spoken. Markets of this type, in addition to providing local agricultural products—vegetables, meats, and cheeses—serve as important social arenas that reinforce long-standing relations between sellers and customers. They can even serve as a central place for candidates to make their cases during local elections. Unfortunately our visit to Céret took place when the market was held elsewhere and we were unable to confirm that Catalan remained a common means of communication there.

Céret is well known among art lovers and intelligent tourists in France for its unique municipal museum of modern art that houses a collection of works donated by a group of first class artists who worked there in different periods and for different lengths of time from the beginning of the twentieth century to the 1960s. Pablo Picasso was perhaps among the most famous but, as Céret's reputation as an artistic center grew, other painters and sculptors who were to distinguish themselves came to spend various periods in the company of their contemporaries and to benefit from the congenial atmosphere and reasonable rents in the area. During this local renaissance, the painters Soutine, Dalí, Chagal, Juan Gris, Miró, Matisse, Tapies, and the sculptors, Manolo (a friend of Picasso), and the Catalan, Maillol, all spent some time there. During our short stay we visited the rich collection in this museum, recently housed in a fine modern building. Céret almost lost its collection when, in the 1990s, the extreme right-wing National Front won the mayoral election. The first action of the new mayor was to fire the museum's director to replace her with a conservative hack who, he felt, would cleanse the museum of its collection of "degenerate" art. The response of those artists still living who had contributed their works was to ask that their donations be returned to them. Fortunately for the town, as well as art lovers in general, the election was contested on the basis of possible fraud. When the issue was brought before the court, the election was declared illegal, and the former mayor was restored to office. He reappointed the former curator and the museum continues as a prime example of the municipality's cultural commitment. It is also an economic draw for many tourists who come specifically to visit the internationally known collection. After our own viewing of this fine collection, we went into the book shop. There we asked in Catalan whether there were any books available in that language and were pleasantly surprised to be shown a series of well-illustrated, pedagogically sound works for children relating to the museum's collection, all in Catalan. We were also pleasantly surprised to find that the clerk on duty was fluent in Catalan and not at all shy to speak it with us. At last we had met someone in Céret who spoke Catalan and was proud of it.

Two Anecdotes Concerning Self-Hatred and the Catalan Language

As I have already noted, our initial visit to Prades was in the spring of 2000, before we had discovered the UCE. Arriving for the first time we went directly to the tourist office (the *Syndicat d'Initiative*) to obtain a local map. We also wanted to visit the Pau (Pablo) Casals Museum located next to the tourist office, just off the main square of the town. As noted earlier, Pau Casals, world famous musician, cellist and orchestra director, as well as one of the most famous Catalan exiles from the Franco regime, spent most of his last years in Prades where he fathered the internationally famous Music Festival. Even after he left the town for Puerto Rico toward the end of his long life, he would return to direct his beloved festival that, to this day, takes place every summer during the first two weeks of August. Well-known musicians from all over the world come to the festival to play at the nearby monastery of St. Miquel de Cuixà, a wonderful example of typical Catalan religious architecture spectacularly located at the foot of the Canigó Mountain, just outside of Prades.

After our short visit to the museum, we returned to the tourist office to get information about how to find our hotel, located somewhere on the periphery of the town. We were greeted, in French, of course, by a young woman who was very helpful in pointing out our chosen hotel on the map. Then we jumped in and asked innocently: "Do people here still speak Catalan?" She answered angrily, almost shouting at us, with a curt "NO!" About an hour later we returned to the office to ask her about visiting the typical Catalan church that graces the central square of Prades, but which we had found closed. It turned out that, in addition to working in the tourist office, she had a university degree in art history and served as the official guide to the church with its many art works. She agreed to open the church for us and we soon discovered that our guide knew a great deal about the church's architecture as well as its collection of religious art. As she warmed up to us and realized we were not typical tourists, she admitted that during her university studies she had taken an option in the Catalan language in addition to Italian. The more we talked the more she loosened up, eventually admitting that Catalan was the language of her parents' household and, therefore, her first language. She was careful, however, to add that no one in Prades spoke Catalan to outsiders. It was rather, she said, reserved for conversations among relatives in the intimacy of the household! Sadly, Catalan had lost much of its value in the eyes of this native speaker, too embarrassed even to acknowledge openly to strangers that she could speak it.

Later as we asked the owners of local cafés and stores whether they spoke Catalan, all of them replied no and, as it turned out, many, in fact, were originally from non-Catalan-speaking parts of France. A year later, during our first year of attendance at the UCE, we visited the weekly Friday market held around the borders of the town square. It was there that we finally did hear Catalan spoken, but only by a few of the older market folk.

In 2001, we were confident enough with our Catalan to register for the summer session of the UCE in Prades. Before leaving, one of our neighbors in France told us that he knew a person who lived not far from where the university session was to take place and who was of Catalan origin. We phoned the gentleman who agreed to meet us at his house perched on the hills southeast of Prades. His story, as we were to learn later, was not atypical of the experiences that many former Catalan speakers had with their maternal language in comparison to French. This man's father had been a poor immigrant to France where he eked out a living as an agricultural laborer. The father was also an alcoholic. Growing up in this household was a painful experience for our interviewee not only because of the poverty he suffered, but also because of the frequently violent behavior of his father when drunk. Although Catalan was the language of the household, our informant soon came to see French as an escape from his unfortunate environment and as a means of advancement. Throughout his school life, his teachers had told their students that Catalan was an inferior language, associated with backwardness. French, in contrast, provided the means to get ahead in the world. After graduating from high school he went on to the university and eventually became a professional educator. To this day he is thankful for the social and economic advantages that the French language has provided him. When he finished his tale, we sensed a feeling of nostalgia and loss as he told us that while he still could speak Catalan, to do so was a painful experience. Here, as in the above example, Catalan represents the domestic environment one has left behind (or rather, escaped from in these instances) through assimilation of the French language and culture.

A Tale About Naming One's Children in France

When we first visited France in the summer of 1957, a long-standing national law required that the first name all French children had to bear was the standard name of a Catholic saint. This in a country that, since the French revolution, has prided itself on having a strict separation between church and state. At the time major protests against this law were generally limited to Breton parents who wished to give their children typical Gallic names in keeping with their Celtic language and culture. The result of this ongoing protest was that eventually Breton people were granted a special exception to the government rule. From then on they were able to give their children traditional Breton names. In the late 1990s, the European Community passed a law, accepted by the French government, that the naming of children in the European Community was henceforth to be a private matter. Thus, in principle at least, parents in each member country were free from then on to name their children as they wished just so long as a chosen name was not inimical to the well-being of the child or of public order.

During our first session at the UCE, we took a course on the history of Northern Catalunya with Alà Baylac that dealt with local history beginning with the signing of the Treaty of the Pyrénées up to and including contemporary times. The content involved an analysis of French rule, particularly as it impinged upon Catalan language and culture through the years of French domination.

At the end of the course, to illustrate one of the ways the French government defies European law, Baylac brought up a personal matter concerning his own family and the naming of his son. It seems that when the child was born he and his wife chose to give him the Catalan name Martí spelled correctly with an acute accent on the i to indicate the stress on the final vowel. When Baylac attempted to register his son's name with the French authorities in Perpignan he was told that the use of the accent on the letter i was not in conformity with French spelling since such an accent was not a part of the standard orthography of the French language and that, furthermore, French typewriters were not equipped to type this particular sign. This strange response, considering the fact that computers, unlike typewriters, are flexible enough to insert the offending accent with ease on any necessary documents, led the couple to bring a court case against the local authorities in Perpignan. When their case went to trial the decision of the judges found in favor of the city bureaucrats. What follows is an extract from the court's finding on the November 26, 2001:

> All civil acts must be written in the French language, the only official language of the Republic as stated in Article Two of the Constitution.
>
> That the acute accent in French is not used on the letter i, and the use of such an accent on this letter would not modify the pronunciation of the name.
>
> It should be noted that the choice of the name Marti was not refused by the public authority but only the Catalan spelling of this name.
>
> The refusal does not constitute a violation of the European Convention of Human Rights because it conforms to the legislation that imposes the utilization of the Latin alphabet for the publication of civil acts and because the addition of such an accent over the i of the name chosen by the appeal has no effect on the pronunciation of this name in the French language.

The last time I spoke to Mr. Baylac, in the summer of 2003, he informed me that he had lost his appeal to a higher court in France and was preparing to take the case before the European Court of Justice! Meanwhile, while the courts in Perpignan refused to accept the offending accented letter, in another similar case, this one in Montpellier in the Department of the Hérault and therefore not a part of Northern Catalunya, the court found in favor of the parents who had given their son the Catalan name Lluís with the offending accent once again over the i. Here is an extract from that court's decision of October 8, 2000:

> Agreed that while it is true that the Roman alphabet does not recognize the acute accent over the letter i, this accent is a diacritic used in both French and Catalan;
>
> and agreed as well that the European Court considers that the name of a child has an intimate and affective content that is based on the private sphere and that this right is recognized by Article 8 of the European Convention of the Rights of Man;
>
> that the application of the European Convention of the rights of man must take precedence over the Civil Law [of France] as long as the exercise of this right is not contrary to public security, the economic well-being of the country, the defense of order and the prevention of penal law, the protection of health or morality or the protection of the rights and liberty of others;
>
> agreed that this name is not contrary to the interest of the child or of public morality;
>
> THE COURT

orders that the birth certificate of the child Lluís Denis AYMAR ... be corrected to include the acute accent over the i.

Here we see a contrast between the formal, pedantic response of Perpignan and the more humane, tolerant attitude of the Montpellier Court. Could it be that Perpignan feels more threatened by the acceptance of Catalan than the Hérault Court? We might surmise that this is the case.

The French Attitude Toward the Teaching of Catalan in Public Schools

The Valèncian magazine, *El Temps*, in the February 17–23, 2004 issue, carried an article on page thirty-five noting that the same Alà Baylac who is fighting to give his son a Catalan name had informed the magazine that the teaching of Catalan in Northern Catalunya was under a new threat. This from a decision of the French minister of education to drastically reduce the hiring of state-qualified Catalan teachers (holders of the CAPES: *Certificat d'Aptitude au Professorat de l'Enseignement Secondaire*). Until this decision was declared, Northern Catalunya was allowed to hire three professors per year for the teaching of Catalan, a language requested by 60 percent of parents in the region and which, even with the former quota of three teachers per year, was only able to serve 6 percent of the demand. Baylac observed that in comparison with Northern Catalunya, Corsica was allotted fifteen teachers of Corsican. The article goes on to point out that the French government had also reduced the total amount of money available for optional programs within the normal program for passing the standard high school diploma. The article quotes Baylac as saying, "It is shameful [since the decision] severely prejudices the teaching of Catalan and does not respect the promises made last October that the regional languages of France would be protected."

This action on the part of the French government contrasts sharply with the Declaration on the Rights of Persons Belonging to National or Ethnic, Religious, or Linguistic Minorities that was adopted by the General Assembly of the United Nations on December 18, 1992. What follows are a series of extracts from this declaration appropriate to the situation in Northern Catalunya:

Article 1
1. States shall protect the existence and the national or ethnic, cultural, religious and linguistic identity of minorities within their

respective territories and shall encourage conditions for the promotion of that identity.

Article 2

1. Persons belonging to national or ethnic, religious, and linguistic minorities have the right to enjoy their own culture, to profess and practice their own religion, and to use their own language, in private and in public, freely and without interference or any form of discrimination.

Article 4

1. States shall take measures where required to ensure that persons belonging to minorities may exercise fully and effectively all their human rights and fundamental freedoms without any discrimination and in full equality before the law.

2. States shall take measures to create favorable conditions to enable persons belonging to minorities to express their characteristics and to develop their culture, language, religion, traditions, and customs, except where specific practices are in violation of national law and contrary to international standards.

3. States should take appropriate measures so that, wherever possible, persons belonging to minorities may have adequate opportunities to learn their mother tongue or to have instruction in their mother tongue.

In 1992, the Council of Europe passed the Charter for Regional or Minority Languages. What follows are important sections of that document:

Article 8—Education

1. With regard to education, the Parties undertake, within the territory in which such languages are used, according to the situation of each of these languages, and without prejudice to the teaching of the official language(s) of the State:
    a.
    i. to make available pre-school education in the relevant regional or minority languages; or
    ii. to make available a substantial part of pre-school education in the relevant regional or minority languages; or
    iii. to apply one of the measures provided for under i and ii above at least to those pupils whose families so request and whose number is considered sufficient; or
    iv. if the public authorities have no direct competence in the field of pre-school education, to favor and/or encourage the application of the measures referred to under i to iii above.

Sections b, c, d, e, and f. of the same document repeat these requirements for primary, secondary, vocational, university, and adult education.

Section g and h require

> arrangements to ensure the teaching of history and the culture which is reflected by the regional or minority language and to provide the basic and further training of the teachers required to implement those of paragraphs a to g accepted by the Party.

I now ask the reader to contrast the above articles of the General Assembly of the United Nations and the Council of Europe to the following declaration by Claude Allègre, well-known French scientist and former minister of education in the national government.

The French magazine *L'Express* for September 8, 2001 carried an article by Allègre on page twenty-nine. The entire piece is a defense of French against the teaching of regional languages in the national elementary and secondary school system. The title of the article, *Gardarem Lou Frances* (Save French) is, perhaps for the sake of irony, in Occitan, the childhood language of the author. Allow me to quote a few parts of it:

> In the not so long ago past my grandmother was struck across her fingers with a ruler by her elementary school teacher if she were caught speaking Occitan to her neighbor in the classroom.... This is how we made France! When I hear the extollers of regional languages plead for a Europe of regions, I think that their program is to definitely destroy France! In an uncertain world, any weakening of France is a weakening of Europe....
>
> The teaching and practice of regional languages should be organized [exclusively] in research centers of research universities. Their teaching should be encouraged and financed by those who wish to reclaim them, that is to say, the regions....
>
> Occitan is a real language, alive and full of beautiful sounds, a language that, with Frederic Mistral, even had its Nobel Prize. I spoke it very well, but times have changed. Yes, I prefer to develop computer experts in French and to have them speak English instead of the Corsican or Catalan spoken by shepherds.

Some Data on the Current State of Catalan in Northern Catalunya

The most recent study concerning the state of the Catalan language in all the regions it was traditionally spoken is *Viatge a les Entranyes de La Llengua: de la riquesa a l'agonia del Català* by Eugeni Casanova. This book is the result of a long series of recent voyages by the author within the Catalan world during which he sought out Catalan speakers of the language's many local variants to observe the current degree to which it is spoken and by whom. Throughout the work Casanova probes Catalan speakers to reveal their personal relations to the language, the contexts in which they will or will not use it, and their predictions concerning its persistence or potential disappearance.

The initial chapters deal with Northern Catalunya (the Department of the Pyrénées Orientales) and include sections on linguistic usage in the Cerdanya (Cerdagne) Capcir, Conflent, and the more rural parts of Rosselló (Roussillon). The situation he describes is not a rosy one. Everywhere he went he found an alarming absence of linguistic facility among the young. The first chapter opens ironically as the author describes his arrival in the small border town of Palau-de-Cerdagne via the Rue d'Espagne. There in the central square he finds the *tricolor* flag of France "of Aznarian proportions, too large for the building to which it is affixed." [José-María Aznar was the prime minister of Spain from the late 1990s until 2004.] The wall of the building bears a stone tablet in the French language that says "*Aux enfants de Palau-de-Cerdagne morts pour la Patrie*" (1914–1918). Casanova notes that all of these fallen heroes bear typical Catalan names such as Coll, Colomer, Forn, Lacaza, Maurell, Salas, Sans, Orriols, and asks "Who of them were of the *Patrie?*" He goes into the local bar and finds a man who speaks Catalan well but whose vocabulary lacks certain words. Casanova asks, "Who speaks Catalan here?" The answer, which he was to hear over and over again during his extensive travels in the region, was "The old folks. Those of my age know it but don't speak it much and the young don't speak it at all." With few exceptions this was to become the litany in each area of the Department he visited including some of the most isolated parts well up into the mountains.

In his second chapter, Casanova takes note of one of the most ironic facts concerning the state of Catalan in Northern Catalunya: The population that uses it the most in their daily lives is the Gypsy community of Perpignan. Casanova informs us that the section of the city where the community lives is the only location of unilingual speakers of Catalan in the region. Here the language is so well anchored in the lives of the residents that they refer to it as "Gitano" rather than Catalan.

Casanova devotes an entire chapter (*Parlar Gitano*) to the Gypsy community that extends in lesser numbers beyond the capital city of the Department to parts of the Fenolleda (the Occitan part of the Department) as well as the Vallespir. As is well known, Gypsies are inveterate travelers and Casanova's informants told him that wherever one finds Gypsies elsewhere in France (he mentions the cities of Lyon, Marseille, Carcassonne, Narbonne, and Béziers) who speak Catalan, one can be sure their origins are in Perpignan. Later, Casanova notes that in at least thirty cities of metropolitan France, as far to the north as the frontier with Belgium, one can find Gypsies who are Catalan speakers.

The Fate of Catalan in the North

In his book, *True France* Herman Lebovics describes the development of French cultural identity during the Third Republic both within the various regions of the home territory and in the French Colonies. Of the general policy he says:

> The idea of France [True France] it consecrates is profoundly static and a-historical, indeed antihistorical. For despite all vicissitudes of history—monarchy, republic, empire—a vital core persists to infuse everything and everyone with the undying if seriously threatened national character. (9)
>
> Our conclusion has to be that the construction of an exclusive, unitary, and fundamentalist concept of French cultural identity was a conservative political project. . . . When we reflect on the framing of the discourse of essentialist cultural identity, on the efforts to persuade the French to accept an imperial sense of Frenchness and to persuade the colonial people to return to French-nurtured roots so that becoming more native they could be more French, and finally, on the creation of a domestic cultural definition of what it was to be French which could benefit only the most regressive sides of national life, we find that True France has been deconstructed to reveal its consequences in cultural narrowness and political reaction. (189–190)

What successive French governments were doing, according to Lebovics was to assimilate the internal population of France with its variety of local cultures into the colonial empire at the same time as they were integrating the indigenous peoples into the mother country. Thus the well-known slogan of colonial France—"Our ancestors the Gauls"—was preached regardless of local culture to schoolchildren everywhere in the once far-flung empire. If local languages were at times allowed to persist, as Lebovics suggests, it was with the goal of transforming local cultures into the equivalents of the colonized who were to become essentially French, even as they kept some of their local features. The desired result was far from the idea of a multicultural France.

Instead such languages and cultures were supported, when they were, with the perverse goal of integrating local cultures into the scheme of a united and unitary France. To illustrate this Lebovics cites a poem by Felix Gras:

> I love my village more than your village; I love my province more than your province; I love France more than anything. (175)

Judging from our own postwar experience in France, particularly in the formerly Occitan-speaking part (the Department of the Hérault where my wife and I have had a house since 1968), the rapid decline of local language use through their systematic suppression (or, at the very least, neglect) in the schools, along with the concomitant spread of French as the national language in local areas via national television, was in full swing. This process was accelerated, I believe, by the end of the Second World War and, with the elimination of the Vichy government, to distinguish the restored Republic from the fascist period. As I have already noted, it was not until Jacques Chirac became president of France that a leader of the country was ready to admit that Vichy had indeed been France rather than some extra-planetary aberration. As Lebovics points out, Vichy had fostered local identity as a means of creating national loyalty to its particular conception of France via a concentration on folklore and the peasantry, establishing and maintaining an image of the true France through an ideology of right-wing politics. This ideology was encapsulated on the metal currency issued by the Vichy government. Each coin of the realm carried the slogan "Country, Family, Work" a substitute for "Liberty, Equality, Brotherhood" that had been the standard unifying slogan of French identity of the French Republic since the Revolution of 1789. By the time of Chirac's declaration, France no longer needed to distinguish its "true" identity from the Vichy period. The country had modernized and was now quite homogeneous in culture.

Can the French Left Tolerate Local Languages Today?

April of 2004 saw the election of a leftist government in Spain and a few weeks later the left in France also won an epic victory in regional elections. In Languedoc-Roussillon, the region of which the Pyrénées Orientales is a part, the socialist mayor of Montpellier, George Frèche, was elected to head the regional government replacing the former right-wing leader, Jacques Blanc. A few weeks later, according to Alfons Quintà, writing in the Barcelona daily, *Avui* Pierre Bec a member of the City Council of Perpignan and representative of the UMP (the right-wing party in power on the national level) petitioned Frèche to have the new regional government aid the teaching of Catalan in Roussillon. According to Quintà, Frèche's reply was "Catalan is a patois that interests no one!"

George Frèche is a prominent regional socialist and an astute politician but with a reputation for abrasiveness and some arrogance. One would hope that his attitude is not representative but, unfortunately, as head of the regional government, it is important to have his support.

The negative attitude toward Catalan (as well as other "regional" languages in France) was confirmed shortly after Frèche's declaration. In March of 2004 Spain elected the socialist, José Luis Rodríguez Zapatero, as its new head of state. Among his earliest declarations, effectively reversing the position of the former right-wing Aznar government, was Zapatero's tentative willingness to present Catalan (which, after all, is one of the four official languages of Spain) to the European Union for acceptance as an official language of the Community. France, as one might expect, immediately objected to this inclusion and unfortunately France's position was then seconded by the government of the Irish Republic.

Chapter Seven

Northern Catalans Speak

The following interviews are with a small number of Northern Catalans. Four of them are important militants involved in different aspects of Catalan culture. The fifth, although born in the *Principat* and by birth a Catalan living with her French-Catalan husband in the north, sees herself as both Catalan and French or, perhaps more accurately, as Catalan in culture and French in nationality.

Interviewee 1. Man, sixty-one years old. Catalan activist, member of a Catalan association, one of four elected Catalanist members of his municipal council, and the leader of the Northern Catalunya group involved in the transborder festival of Sant Joan (see chapter one). Recorded in French in Collioure, French Catalunya, in 2000.

Before speaking about Northern Catalunya I want to tell you about myself and how I became a Catalan militant. I am a typical product of Jacobin France and the French school system. I was a soldier during the war in Algeria and spent twenty-seven months there on active duty. While in the army I began to understand how the French colonial system worked. Before then I was not at all political. When I came back from Algeria I started to question some of the things that were taught me by my "Mother Country." I had come to realize that there were a number of lies taught to us in school and decided to reconstruct our history on my own. Since then history has been a major interest of mine. I wanted to know what really had happened here in relation to what we were taught and what was left out. For example, we were taught about Joan of Arc but nothing about our own [Catalan] heroes. Little by little I came to understand that all the history we were taught concerned France with nothing about the regions of the country; how they differed and how they were incorporated into the nation. About the same time, I learned that across our border to the south the people were not Spanish. They didn't speak Castilian but, rather, the same local language as we did. [Are your ancestors from here or from the south?] I am half and half. On my father's and his father's side, I am from here. At home, I spoke Catalan with both my parents as well as with all my grandparents. Even in the 1960s most of the people spoke

Catalan. We considered people from regions of France foreigners, not in a pejorative way, but rather to distinguish ourselves from them. Generally speaking, if one went into a butcher's or a baker's shop the conversation would immediately stop until people realized what language was being spoken. Nonetheless, the majority attitude here was that French was the dominant language. In school we were told to speak French and not Catalan. I remember that one of our teachers had the good sense to use Catalan while teaching us French, but he was an exception. The official position was to forbid Catalan. The only written access we had to Catalan was in books published on the other side of the border, but most of these concerned religious subjects. They came from the Abbey of Montserrat where Franco tolerated the publication of certain Catalan works (mostly religious tracts). There was an agreement between Franco's government and the church concerning this issue, because Franco presented himself as a guardian of the religion. He was probably unaware that there was a great deal of anti-government activity at the Abbey. Among the books published by the monks a number of them were pro-Catalan and anti-Franco tracts. These were carried out of the monastery under the hassocks of the priests to be distributed around the *Principat*.

During this time the provisional government of Catalunya was in Toulouse, the parliament in Mexico, the national council of the Catalan resistance in London, and the president of Catalunya had his office near Paris. All members of this Diaspora hoped to return one day to Catalunya. From time to time the members of these different groups met in secret in Sarrià de Ter [in the *Principat*]. After my political views had changed in 1973, I began to militate for a political party we named the *Partit Federalista Europeu de Catalunya*. This party was recognized by the Federalist International, of which the majority were from Germany and Switzerland where the political systems were already federalized. The party was created between 1971 and 1973. Beginning in 1973, I began to attend meetings on the other side of the border, an activity that was strictly forbidden by the Spanish government. At the same time I discovered that an enormous number of things had been done to destroy our identity here. This is my reproach against the French political and educational systems that have consistently attempted to level everyone. Concerning Catalan, for example, the French said we did not have a language but only spoke a patois. Dialect is a more polite term, but it means the same thing. The word "patois" was invented at the time of the French revolution. [What effect did this have on the feelings of Catalan speakers?] Simply put, shame and inferiority in relation to ourselves.

According to the French, but also to many of us as well, one had to speak French to be someone! In the schools before 1940, there were posters that said: "Be clean speak French." That is to say, the Catalans were pigs and, of course, it was the same thing in Brittany, and other regions where French was not the local language. I remember another poster that stated; "It's forbidden to spit on the ground and speak Catalan." These slogans, and others like them, were generally printed within a blue-white-red border, the colors of the French flag.

It was also the case that to get ahead in one's profession you had to begin your career in Paris. If you achieved the middle-school diploma you were told you could begin your career in Paris as a mailman or, perhaps, work for the railroad. Or, if you had the high-school diploma, you could begin in Paris in a more advanced administrative post. If you wished to return to your own region you had to follow a prescribed route through other parts of France. Catalans, Alsatians, Bretons, and Corsicans were assigned by seniority rather than by their home territory. This system began to change only in the 1960s when regional governments were created. Since then recruitment has, in effect, taken place within the regions. However, much more progress is needed to attain real regional autonomy. Most of the railroad workers here, for example, are still from other regions. A third of the tax officers here are *pieds noirs* and, I must add, another third are from elsewhere outside of the region. I'm not against these people as people, but use the example to illustrate how the system continues to operate. This is how the government created the present situation in which Catalan is a minority language. A great number of people here regret this fact but do not go so far as to actively reject it. The general idea now is that it's too late to do anything, so such people refuse to fight for their rights. There are very few people like me who are militants in a political party that defends Catalan culture and language. Many will say in private that they are fed up with the situation but don't want to commit themselves to fight for their rights. I am, perhaps, an exception because I developed many of my militant ideas in Paris [where he worked for a good part of his active life]. I learned to confront the authorities, for example, in strikes. During my years in Paris I could have accepted a transfer to Marseille, or Toulouse, but they would never have transferred me here. The political party to which I belong works for the kind of autonomy that already exists in the *Principat*. Overall we are interested in the development of a Europe divided into "Euro-regions." Our Euro-region would consist of the Midi-Pyrénées, Languedoc-Roussillon (of which Northern Catalunya is a part) and the *Principat*. It would therefore be an across-the-border political unit.

We include the non-Catalan departments in Languedoc-Roussillon in our hoped-for region because, as Occitans, the people there are close to us in culture and, therefore, their presence would not break the coherence of the unit.

[What holds your culture together? Is it mainly the language?] The language is the major factor that holds a culture together. For the schools here the city council has created a Catalan program, half of which is paid by the *Generalitat* and half by the council. This does not mean a lot of money, but it does support three to four hours of Catalan language teaching each week for each class in the elementary school. We have had this program for the last four years. There are some other towns in the region that have also begun this type of program. Those who teach Catalan in the schools are not necessarily Catalan in origin, but they must, of course, be familiar with the language and the culture. We also have the *Arrels* schools [*Arrels* means *roots* in Catalan]. These were originally private and took students up to the age of twelve. As private schools the entire program was in Catalan. Unfortunately there was a conflict with the government concerning the *Arrels* and, as a result, they were integrated into the public school system but only on condition that they include the teaching of French in their required curriculum. Now there are one hundred seventy students in the *Arrels*. There are also the *Bressoles*, private schools that receive some aid from the *Generalitat*. Almost a hundred percent of the students in these schools succeed, while in the regular French schools the success rate is much lower. Ironically, perhaps, in the beginning our schools were accused of being inferior to the regular public schools, but now, because they are successful, they say we are elitist and refuse to choose the lowest performing students! The fact is, however, that Catalan schools admit students whose parents are looking for an alternate education for their children, rather than because their children are particularly intelligent. Most of these parents are young people and not all of them speak Catalan. So far there are only a few schools of this type, public or private. There are five *Arrels* and [at the time of the interview] three or four *Bressoles*. There is also another public school system with programs in bilingual education that teach half in Catalan and half in French. These began in Perpignan and now have about three hundred students. During one year mathematics may be in Catalan and in another year geography, for example. There are schools with this program in Céret, Prades, and Perpignan. This is an equitable system. We are not against other languages. Many Catalans are, in fact, trilingual in Catalan, Castilian, and French.

The University of Perpignan also has a Catalan department, and some of the graduates in this program become militants. There are also adult courses in Catalan in several villages around the region. Even some foreigners take the language in these classes. There are also French people from outside of Northern Catalunya who now live here and want to learn Catalan. For the last two years it has been possible to sit for an International Certificate in Catalan at Perpignan, although the diploma is actually awarded by the *Generalitat* in Barcelona. The examinations for this degree are sent to Barcelona where they are graded. Last year there were two hundred forty-eight adult students in this program and, of these, two hundred forty passed. Many of the people who took the course already spoke the language well but wanted a diploma. This year there were two hundred ninety-eight candidates. Twenty-five percent of these students failed. What is encouraging is that the number of those who took the exam has grown. Some of them were young people who felt that the diploma would help them to find employment in the Barcelona area.

Every year for the festival of Sant Jordi [Saint George] the Catalan national saint's day, there is a lecture in Collioure in Catalan as well as an exhibit related to the theme of the lecture. On this day everything takes place in the Catalan language. We don't allow any translations during this celebration because this is a special Catalan day. We do, however, make it plain on the posters advertising this event that it is exclusively in Catalan. We refuse to accept the fact that if even one person is present who doesn't understand Catalan, we must provide a translation. I remember that one time a Spaniard came and demanded that it be in Castilian. We, of course, refused. [What is the percentage of people who speak Catalan here?] It's a population pyramid that's upside down. There are many old people who speak the language but very few of the young do. In the sixties we had ninety percent Catalan speakers. In the eighties it dropped to seventy percent. In 1990 it was fifty percent and now it has fallen to forty percent. [What about those who know how to write the language?] Ah—that's much lower, but now it's beginning to rise again given the various courses that are available as well as the Internet. [What is the situation in other parts of Northern Catalunya?] The more people from the outside who come here as vacationers or as settlers, the less the Catalan tradition is kept alive. The further up you go into the mountains, which are the most isolated regions, the more you will find the tradition alive. Here the Sant Jordi festival has once again become popular perhaps because, in recent years, it has become an international celebration. [In the Catalan tradition on April twenty-third each year, men present a rose to their wives or women friends and women present books in return to their men. Now-a-days the celebration involves the exchange of books and roses between good friends of either sex.]

I continue to spell my name in Catalan rather than in French. After all, one's name is a strong part of one's identity. [If someone asks you your nationality what do you reply?] Catalan! [His wife interjects.] It's always been that way, even in Franco's time. When someone asks him if he's French he replies—"No I'm Catalan!" When he crosses the frontier the first thing he shows is his Catalan identity card [which, of course is, not an official document]. [He adds:] Even when I cross an international border I show my Catalan card.

[How do you feel about the European Community's law concerning regional languages?] I have signed petitions demanding that France sign the entire law rather than objecting to certain parts of it. This law is based on a spirit of cooperation among those who participated in the writing, and France was one of the participants. When it became obvious that the law's passage could not be stopped, the French parliament passed a constitutional amendment defining French as the only official language of the Republic. [This contrasts with the constitutions of Spain, Finland, Belgium, Luxemburg, and Switzerland, all of which, except Switzerland, are, like France, Members of the European Community.] Three months later France was condemned for this behavior by the members of the Community. Our own response to the French government was to organize protests including hunger strikes. We were joined by other regions of France that share our linguistic concerns. We also wrote to the presidents of France (first to Mitterrand and then to Chirac) and each time the response was that they could not support certain parts of the European Language Charter expressly because it was contrary to the French constitution! But it was they who invented this contradiction. France has taken a hypocritical stand on this issue. There are many sections for each article in the Charter. Of the one hundred sections, France has chosen only thirty-nine, the minimum required to be in accord with it! We wanted France to sign all of it because we knew that if it didn't, the law would be useless to us in our fight to preserve our language rights. Even Russia has said that it was willing to sign the entire Charter.

When New Caledonians or Polynesians defend their language, I feel that they are also defending mine. Dignity is for everyone or for no one. As I said, I am not anti-French, but when they deny freedom to others I protest this attitude.

Interviewee 2. Man, sixty years old. Retired, but active in support of the language and culture of Northern Catalunya. He has also worked hard to insure the success of the UCE from almost the beginning of its creation in the 1960s. Recorded in French in a village near Prades in 2000. The interview began with the following statement by the interviewee:

You say nonviolent, but if the Indians in America had been nonviolent they would also all be dead. There are people that at certain times must utilize nonviolence when they have no other means of protest. Catalans are less violent than, for example, the Basques and the Corsicans. And we will die out anyway. You should be aware of one thing. It's not a question of political parties in France. There are as many people on the left, perhaps as many as on the right of the political spectrum, who practice this genocide. The system that liquidates cultural differences in France finds its support everywhere. Here, they do not liquidate us with arms, but with what we wear, the radio, the TV, even what we eat. There are people who want to aid us in our struggle, but every time we take one step toward progress the system takes fifteen steps against us. The system crushes us, but not only from within France. The Germans fought three wars against the French and now they are waging a fourth; now the weapon is money. There is no defense against this economic war that, I must add, is also supported by the American dollar. One must be lucid about all of this. Our struggle is a hopeless fight between a clay pot and an iron pot and we are the clay pot!

[You were one of the first involved in the creation of UCE?] Yes, but that was a long time ago. The university has existed for thirty-two years. Imagine what it has taken to continue. But looking at our general success we must also ask "how many people have we convinced?—forty-five intellectuals and twenty-three foreigners?" That's the problem. If one is an optimist you can say it's progress but if one is a realist you know we're finished. I'm surprised that two Americans are interested in Catalans, but that's only two out of two hundred million. Take the example of the village of Eus [near Prades]. It is an architectural wonder. What has happened to it? Strangers from outside the region have bought land there in order to construct houses where they can profit from their beautiful view of the village. At the same time what they have built is foreign to local architecture and ugly.

[What about the Catalan schools? Does that not provide some hope for Catalan culture and language?] I already told you if you are an optimist you can say that the schools are an example of progress, but I can't agree. The Catalan schools *are* fine but serve only a small minority. Take another example M—— from —— gives six public lectures about

Catalunya every year in his local region but, on every occasion his audience is small, around thirty to thirty-five people. Of these, at least half are foreigners. Perhaps some Polish people attend who are interested in Catalunya and even speak Catalan. Magnificent! [What about the Catalan courses for adults?] They are what they are. Listen. It's as if I went to America to see the Indians on their reservations. I'd learn their language to better understand them but all the reservations should be disbanded! There are people who go to these courses to learn to speak Catalan but they can neither read nor write it. There are twenty-five to thirty people here who take the courses, but even if there are similar courses in thirty or forty other villages how many people does that make! The population of the department is three hundred and fifty thousand and of them only one hundred and fifty thousand are Catalan. That means that a newspaper in Catalan here might sell as many as eight hundred copies, but rarely more. There are a thousand people here who buy a book in Catalan each year. The south [across the border] has another dynamic. [Do you have relations with the people from the south?] Yes, of course. [What are these like?] They are mostly cultural. We visit there frequently, particularly Girona and Barcelona. People from the south come here to visit us as well. [What was the language of your family?] Catalan! There are some young people here who do speak Catalan. My daughter went to a Catalan school. [We tell him that we sympathize with Catalans for many reasons, including the fact that we have both lost the languages of our parents.] Here Catalan, if it does have a chance, has it because of the south. Without the south it's finished in the north. Occitan is already dead to the east of here! And, speaking of the south, it's necessary for the people there to remain highly vigilant if they don't want to lose their language. They live in a country that is very centralized, very right wing. Even the left wing, for example, the Socialist Party, is very centralist. [What does the French language mean to you?] That's a very complicated question. French is not my enemy. My enemy is the people who use French in a certain fashion. In fact I wish that I could also speak Italian, and English, etc. That's not the problem. The problem is not that we were educated in French, but rather in the content of what people taught us in that language. You can say something in French that pleases me and something else in Catalan that does not please me at all—you understand. What I object to is how the French have inculcated certain values in us. They have tried to make us think that it's French and nothing else. They used to say, "Wash your hands and speak French and if you spit on the ground, you speak Catalan." That is what has made us so angry. On the other hand, to learn French in class and to be able to understand the poetry of François Villon—I *am* for that. But instead of teaching us Villon they told us not to spit on the ground and not to speak Catalan. The educational system in France is where the combat belongs. It's a weapon with two cutting edges. From the point of view of theoretical democracy they say "obligatory education, free, and secular." But one eventually comes to realize that the French school system is neither obligatory, nor free, nor secular. Those who do not want to go to these schools and suffer an education of this type must go to private schools where they have to pay. Therefore, in the long run, even those in public schools have to pay the price of poverty. It's not secular because there is a system of indoctrination that is close to a religion. The teachers are priests of secularism. They preach equality of opportunity, but it's not true. The people don't have equality of opportunity. A child who comes from a family in which the father is a doctor or lives in a house that has a collection of books has a greater chance to succeed than a child who comes from a working-class family. [But are there not some of the latter who by their education, can improve their social situation?] Yes, but that's just an alibi. The French educational system has led us to believe that French is the only language of social advancement. This was originally an agricultural region. For those who could not become independent farmers the government said, "You can be a functionary, but you must serve elsewhere in the nation, and, therefore, to get ahead you must speak French." People who worked outside of their region were granted several years of seniority over those who never left. The system rewards these workers for going elsewhere. I am from the generation that did have the opportunity to learn oral Catalan as an elective in school, but the class was for only one hour per week and we were not taught to read or write. That was in the 1950s. Then we were the first generation to teach Catalan but without a diploma in the language. In 1974 we imposed writing and reading on our students in Catalan classes. This was in middle school. Before that only children under the age of five in nursery school were allowed to learn some Catalan. It was the Basques who first won the right to teach all subjects in their language in elementary and then in high school. Here Catalan schooling after the age of five had to be private. When these schools began, lay people accused them of being religious schools. But they were not religious schools at all. The language is now taught through high school, and there is a program in the university that gives advanced degrees in Catalan. But on the elementary and secondary level all the schools together have only about two thousand children among the total school population of one hundred thousand. We have only one Catalan radio station here and the others broadcast very little in the language. We have *El Punt* here [a special issue of a Catalan language newspaper published in various regional editions primarily south of the border]. It is the only newspaper published in Catalan that is readily available throughout the region. Very few copies of *Avui* are sold here, perhaps one hundred or one hundred fifty copies in the larger urban centers. Here everyone reads the [French language] *Indépendent*, and politically that paper goes with whatever direction the wind takes at the time. [What role does Perpignan play in impeding or fostering Catalan culture?] The current municipal council has made an effort to open Catalan culture as an image for the city. They pay a small amount for education in Catalan and I can't say there is any negative aspect to their policy, however minimal it is. It's a right-wing municipality. But right or left it would be the same. In the south the national Spanish government cannot bypass Catalunya. There the Catalans have enough weight to negotiate with the central government. Here we have no weight. [What about decentralization?] What decentralization! The only place to get autonomy among French territories so far has been New Caledonia. Even the Corsicans have not arrived at that stage yet. If Corsica does win autonomy sometime in the future perhaps there will be some hope for us, but the word that the French understand the least is "autonomy." The word is square and does not fit into a hexagon! Chevènement [The former minister of the Interior under the socialist government in the 1990s] said there were two words he did not want to hear. They were "autonomy" and "Catalan." It was Catalans from the *Principat* who went to Corsica on the Corsicans' request to explain how their autonomy worked. This resulted in the Corsicans telling Paris that they wanted the Catalan type of autonomy for themselves. I don't believe that Northern Catalunya will ever have autonomy. One reason is that there are only one hundred and fifty thousand Catalans here. The neo-Catalans [those born elsewhere but who have to varying extents accommodated to the local culture] are sympathetic but they are not Catalans. Some of them send their children to the Catalan schools because they don't want them to go to school with Gypsies or Arabs. The neo-Catalans are people who are fleeing from something. The Southern Catalans up to now have had the power to assimilate immigrant populations. This is a power we no longer have. [Can we return to the UCE and the effect it has here on the struggle for Catalan culture?] The school has not changed much since the beginning except for the fact that it grew very rapidly. On the other hand, the motivation behind the school is now different than it was just after Franco's death in 1975. In the early days the people who came were more politically motivated than they are today. People from the south came because they could express themselves freely which was impossible to do at home. [What about the program as it now exists?] There were eighty sections at the UCE this year.

[At the time of this interview we had not yet attended the UCE.] You can read different things into what happened depending on what the person who writes about it saw and did not see. There are people who spend a few hours in a class and who then go to a workshop and skip the other events. People who attend this way do not get a global view of the entire program. [Tell us something about the people who work for the UCE in various capacities?] During the session the support staff is composed of about twenty-five people, some very young and some older. We have a nursery for those who work and have children, and most of the teachers who work in it are volunteers. We have heard that there are some people involved in the school who want to change it's location. [Can you tell us about that?] It has been suggested by some that it takes place in a different Catalan area every summer, for example, one year in València, one year in the *Principat*, etc.

[How is Catalan culture, beyond the language, expressed in Northern Catalunya?] Whenever there is an expression of Catalan culture in a sporting event, for example, you only see Catalan flags in the crowd, but no one would ever burn a French flag. There are many people who speak Catalan but are afraid to speak up for the language if they are right-wingers or who are Jacobin even if, by chance, they are on the left. [Can you say more about this negative attitude?] All the regions of France have a history of formerly strong cultures. The villages around the country have a monument, usually located in a prominent place, dedicated to the dead of the First World War. Most of these "heroes" were people who, when they were alive, were kicked in the behind by their own government and many of them knew it. But now that they are dead it is necessary to make them into good French men. Here there were two thousand members of the underground during the Occupation out of three hundred thousand in the total population. The others either did nothing or were collaborators. But as soon as the war was over—everyone claimed to have been in the resistance. Likewise, if the Catalans here win their rights sometime in the future, then the people will all be Catalans! If they lose it's another thing. The people have been conditioned by the system. For example, most of the wealthier people—doctors, lawyers, notaries, and druggists—have real antique furniture. When the antique dealer comes to the door he offers them formica "free of charge" for their valuable objects! Once people are conditioned this way you can do what you want with them. [What is it possible to do?] We have to penetrate the parties. For us that means the left since it's impossible to penetrate the right. But we have no real access to the radio or the TV. Our only hope is that the south progresses more. Some of the young who visit the south or work there come back and describe the good things they have seen. That, of course, provides some hope for Catalan culture here. But there is another problem. In each of our schools there is an advisory council made up of parents. If one single parent says no to a project, teaching Catalan, for example—it's no! So if the majority wants Catalan taught, and one parent disagrees, they can't offer it.

Interviewee 3. Man, in his fifties, professor in the Department of Catalan Studies at the University of Perpignan. (Interviewed in Catalan at the UCE in 2002.)

[Is your family from Northern Catalunya or from the south?] I have no relatives in the south. I am a pure Northern Catalan.

The border, I must add, is very permeable. There are many people with relations on both sides. Although my mother is from the north her parents and grandparents all spoke Catalan and, with considerable difficulty, French. My parents were both working-class people. My mother was a cashier in a bank and had to learn French for her job. A great many people here learned French because of their professional responsibilities. French has always had social prestige here. [What about your father's family?] My father is from a non-Catalan speaking province of France. His mother moved to Perpignan when my father was very young. As late as the Second World War the social ambience in Perpignan was totally Catalan. A person from the outside who wanted to work there had to learn Catalan. Of course the rich or well-off passed over into French. My parents always spoke French with me. I am part of the postwar generation that assimilated voluntarily. It was a collective phenomenon. At that time people who were Catalan speakers at home began to speak to their children in French because it was seen as the language of opportunity and economic advancement. My father was a mason, a member of the working class, and we were a family of modest means. I was a very good student from the beginning so my family made every effort to help me succeed. This was general for most children during the same period. Catalan rapidly began to lose all its prestige. It was seen as the language of farmers or refugees from the Franco war, as well as Catalans who immigrated to France from Spain in the fifties to escape poverty in Spain. All these facts seemed to indicate that Catalan had no future, and French was the door to success. In the fifties, sixties, and seventies, there was great social and economic progress in France. For example, my brothers succeeded in becoming schoolteachers and I teach at the university. Obligatory education in French is the basis of the French Republic. [Why then did you decide to perfect your own Catalan?] There was a moment when a whole generation began to return to the language.

Not everyone, of course, but many people, almost a whole generation. My friends all spoke it. In the beginning of this return to Catalan in Northern Catalunya, during the fifties, sixties, and seventies, all the people who spoke Catalan mixed it with French. They lived in a situation of flexibility in which, for example, they would use French in half a sentence and Catalan in the other. People were tolerant of this. Now this is no longer the case. Moreover, the overall ambience had remained Catalan because there was still very little internal French immigration to this region. The ambience of village festivals and local sports events was still very Catalan. This began to change with an increase of migration from other parts of France of people who tended to be very intolerant toward the speaking of Catalan. They would often say, "We are in France, why speak Catalan?" When, in the seventies, for example, they began to teach Catalan in the schools and then, later, in the university, many people who were from outside the area complained about it. By 1968 the situation had become critical. Those of us who cared about Catalan began to fear the growing flood of outsiders who were increasingly critical of us. Also, during this time, we were not really aware of what was happening on the other side of the border. Finally, when travel to Spain began to pick up, people from here would report that Catalan was forbidden below the border. Also, in the late 1960s, the *nova cançó* [Catalan nationalist or Anti-Franco sentiments expressed in song] tied to the post-1968 radical movement began to develop. Although Franco prohibited singers of this movement to perform in most of the south, they did come here to sing. These facts taken together had a strong effect on nationalist feelings in our region to the point where many people became militants. "I am French but also Catalan" they would say. My own change of heart began during 1964 and 1965, when I was a school-teacher in a small local school. It was while preparing the content of my classes I came to realize that the culture and history of my own region were very little known. I also learned that there were many things that had been badly taught as well. Important events in local history were ignored because, for the centralized educational system, these were not considered part of the history of France. These discoveries all contributed to raising my consciousness. When I got my first university degree at Montpellier in 1966 and my *agrégation* in 1968 [the *agrégation* is an advanced competitive examination the passing of which allows one to teach in the French high school and university system], there were very few students in my specialty, and even less who had passed the *agrégation* so I was immediately offered a university post. My first task was to assist a professor of history at the University of Montpellier. By then I saw myself as a Catalan who was an expert in my subject rather than an outsider who was merely interested in Catalan culture. My studies for the *agrégation* taught me the methods necessary for researching the things left out by the official history books. During these years of work my feelings concerning the necessity to preserve Catalan culture took their full form. When I became a professor in Perpignan the head of my department suggested that I teach my courses in Catalan. Before that, although I spoke Catalan and was never ashamed to use it in public, I felt blocked from using it in university teaching.

In 1973–1974, a program in Catalan studies was founded at the University of Perpignan. In 1977 we decided to offer a diploma in Catalan studies. All the courses in this program were taught in the language. Non-Catalan students accepted in our courses were not obliged to speak Catalan in class but had to listen and understand the language. The program required all students to take a final examination and write an essay of about fifteen pages in the language. This was an embryo for offering a full program in the language. In 1968 with the help of relations we had established with the UCE and other competent people, we were able to invite top quality teachers to lecture in the program. It was not long before it became popular and the lecture halls rapidly filled with students. In 1981, we created the Center for Catalan studies. [Did you also participate in the foundation of the UCE?] No, but I joined it in 1974. I gave five or six lectures that year. After that I participated from time to time, but not on a regular basis until 1980. Between that date and 1990, I regularly gave one course every summer. When there was a change in the old system in 1984 my group took over the secretariat of the UCE. Before that date there was no single patron of the school. Each Catalan-speaking area had its own cultural entity involved, including La Franja de Ponent [the Catalan-speaking part of the province of Aragon] and Andorra. Under the new structure, the Institute of Catalan Studies in Barcelona took over the direction with the help of a group of nonaffiliated individuals. I joined as well. [How do you feel about the evolution of the UCE?] There is a constant evolution. In 1984 there was a strong change and then a second, less-radical change occurred in 1996 when control passed to a foundation. The foundation provided a more stable structure that, because it is better regulated, generates more confidence if some crisis occurs. Now, every two years or so, there are orderly changes in the themes and contents of the course offerings. The directors of the foundation decide which changes to institute. Last year there was an attempt once again on the part of the president of the Institute for Catalan Studies, joined by a group of university professors, to change the structure of the UCE. Their stated goal was to make it more academic but, of course, it is, and has always been, academic. They really meant something else. They wanted to have greater control over the school but there was a strong reaction against this change. The structure that the UCE traditionally followed was that the patrons would designate the president who was the person responsible for making decisions concerning the program and who was to teach in it. The Institute for Catalan Studies wanted the patrons to have more power in this area in order to gain control of the academic structure. [Does the UCE influence the *Principat?* and what are its effects on Prades?] Regarding the latter question, there is very little direct effect, the people of Prades mainly ignore the UCE. They are aware that it exists and that it's Catalan, but they really don't pay attention to what happens at the school. Beyond Prades, in Northern Catalunya, in general, they also know about the UCE. All the same, people are generally ignorant about the way it functions. Very few inhabitants of Northern Catalunya take courses at the UCE; only about one hundred per year. But, in other ways, the school does have a notable impact. It generates a significant amount of reflection about the relationships between Northern Catalunya and the *Principat*: a type of reflection that does not occur in Barcelona. The UCE also indirectly provides an important image of Catalan culture for the people of the north as it gives people on both sides of the border some idea that things *are* happening here concerning Catalan culture. They know that the UCE provides a network of contact between the different parts of Catalunya. Also thanks to the UCE, politicians in the north have met their homologues from the south who participate in the sessions of the UCE. [What is the economic relationship between the UCE and the town of Prades?] The businesses in Prades certainly profit to some extent from the students at the UCE. Nevertheless, let us not exaggerate. It's clear that this is not a major impetus for change, but, at the same time, it does help to improve the image of the school. [How is the UCE supported, given that tuition as well as room and board are certainly below cost?] The UCE gets no direct support from the *Generalitat*, and many people participate free of charge. However, some cities in the south as well as private organizations help. Perpignan, for example, supports certain things. The Institute of Catalan Studies that does so much in other areas of Catalan culture and, therefore, has high expenses, receives very little support from the government so that it cannot subsidize the UCE.

[Does the fact that Catalan is the official language of the *Principat* influence people here?] Yes, absolutely but, again, not directly. Barcelona does give some support to the *Bressoles*, but, beyond that, they do not support other activities in the north. Nevertheless, the people know that if one learns the language they can find work on the other side of the border. But the fact that there is the *Generalitat* and that Jordi Pujol is the president [now the former president] and the fact that the politicians here are obliged to keep in contact with him and the government are quite important. [Does the openness of Europe support the development of language here?] Directly, no. The fact is that Europe exists in its present structure as a coalition of states. If a country does not accept a decision of the European Community, the Community has little or no power to impose it. France, for example, has been able to escape major parts of the language law passed by the majority of the Community because it refuses to ratify parts of it. In this respect Europe is unable to facilitate a change toward the recognition of Catalan as an official language. The situation is blocked. On the positive side Europe helps indirectly because of the elimination of the frontiers among states. In line with this change the European Community gives financial support to interchanges among communities that lie on the border between states. Also the fact that people can work in any state of the Community without formal permission is very important. Labor migration of this type increases communication among people of different states. It obliges the development of multilingualism whether or not it's wanted. This is important but not ineluctable. The present dynamic does help, but if there are no actions taken to encourage the phenomenon it will die out. It's a good instrument, but not in and of itself. If the mayors from both sides of the border come together to create joint projects, as some of them have indicated they would, that would be very important. When such interchanges are proposed, even when certain mayors from the north find it difficult to speak Catalan, they will come because they want to participate in such cross-frontier projects. These interchanges function very well to cement cross-border relations. This is typical of the kind of evolution that Europe is able to provoke. Europe organizes and pays for a program known as "INTEREG"—which stands for Inter-Regional Planning. INTEREG is a European project that gives money specifically for transborder activities. The monastery of Saint Michel de Cuixà, near Prades, for example, was given a million two hundred thousand euros by the European Community. INTEREG, however, does not finance cultural or educational projects like the UCE or other scholarly interchanges. But it does support Catalan culture. INTEREG has, for example, subsidized discussions concerning the environment in the Canigó region that lies between the *Principat* and Northern Catalunya. Also, such interchanges take place sometimes in Catalan and sometimes not, although Catalan is preferred since the participants are from Catalan-speaking areas and Occitania. These projects often produce real changes in the participants' points of view.

Interviewee 4. Man, in his early fifties, an administrator of one of the *Bressoles*, the Catalan private schools in Northern Catalunya. Interviewed in English, by preference of the interviewee in 2000 at the *Bressola* in Prades.

[Is there a reason why you wish to remain a group of private schools rather than part of the public school system?] Yes, we do not wish to be state schools because that way you can't have a truly Catalan school. [Can you tell us the difference between the *Bressoles* and the *Arrels* schools?] There is a difference between the old and new schools. In the past both the *Arrels* and the *Bressoles* were old schools, and now this *Bressola* is in the category of new schools. The aim in new schools is not only to teach in Catalan but to create good conditions so that students can speak *and* play normally in the language. That's not true of the *Bressola* in Perpignan and its not true in the *Arrels*. For me there is a difference between these two systems. [But even in Catalunya, schoolchildren often play in Castilian, rather than in Catalan don't they?] That's because fifty percent of the school population in the *Principat* come from Spain. A Castilian speaker in Barcelona can speak Castilian without any problem because of the percentage of Castilians who live there. [Do the children in this school take their Catalan to their home as well as speaking it full time here?] It depends. We do our best to encourage this but can't control anything on the outside. One thing that does help, however, is that we have had Catalan TV here since 1993. Thus the children, if their parents allow it, can watch Catalan TV at home and that's very important for them. [How do you get TV 3 here? Isn't it against the law?] We had to do it illegally as far as French law is concerned, but as far as Europe is concerned it's legal. When they summoned us to appear at the *Prefecture* to tell us that TV 3 is illegal here we replied that *for Europe* it's legal and if you press charges against us you will lose. This is a question of European law against French law and we respect the European law. After that they have not bothered us. There is also a Catalan radio here. In the beginning it was tied to one of the *Arrels'* school radios. This station is legal because it originates in France while TV 3 originates in the *Principat*. [Why do parents send their children to the *Bressoles*?] There are many reasons for that choice and perhaps that's why there is so much demand for places here. This is our best school year so far as the number of students we have, but it's our worst year as well because we did not really have enough room for the new students. In Perpignan they had the same problem and had to start the year in a tent for lack of other space. In the six *Bressoles* that exist today we have four hundred students. [As of this writing there are eight such schools located in different parts of the department.] The plurality of parents really want Catalan but they constitute less than fifty percent of the total if you compare this with the other motivations grouped together. Some parents send their children in order for them to learn another language no matter what language. Some come because they approve of our educational methods. The French schools have excellent *theoretical* pedagogical methods, but they ignore them in practice while, here, we follow the best of these ideas. The problem is that it's impossible to change anything in the French system. The unions tend to block all innovations including those suggested by the minister of education. French education is the second largest business in the world in relation to the number of people it employs. When a minister suggests anything new that we agree with we can immediately adopt it as our practice. And sometimes we do innovative things even before these are thought of by the French government. When it comes to the teaching of regional languages you should consult Jacques Lang's [a former minister of education under the socialist government of François Mitterand] book in which he supports the teaching of regional languages in the French school system, but this is only one example of what I am talking about. Another thing that distinguishes us from other French schools, both public and private, is that we are completely secular. Almost all the other private schools in France are religious in nature. As a result many parents who want their children educated in a private secular school send their children here. [How are the *Bressoles* financed?] First of all we have a sliding scale of tuition. Some families don't pay anything. People with different salaries pay different amounts depending upon how much they earn. As far as state support is concerned each one of our schools has a different relationship with local governing bodies. Here, and in Perpignan, we have the same rights as any French school. The municipalities support certain aspects of our operation. In our school we pay rent to the town government which owns our building. On the other hand they do not help cover the costs of books for our students. Instead they give us a certain sum for each child in the school who is from this local area. This covers about fifty percent of our current school population. We get almost nothing from the department, but the regional government does provide some additional support. [So parents have to make up the difference?] To some degree, but that would not be enough. We try to diversify our sources of funds. We obtain some help from the government of Catalunya, and some from the Balearic Islands. In the latter case the old government of the Islands did not give us any aid. There are one hundred fifty municipalities in the *Principat* that provide some aid. Here, in the north, we receive aid from only ten communities. Additionally, after a private school has existed for five years, the central government in Paris is obliged to contribute to the salaries of full-time teachers in the school. [Do you agree with the common idea that Northern Catalunya as an entity cannot survive without the active financial and moral support of the south?] I think that if you cut off someone's arm the person can live without that arm. And since the foundation of the European Community we have seen a difference in how the Spanish government views Catalan culture here. Additionally, from the beginning, the European Community gave us some financial aid. Recently, however, this type of payment was challenged by one deputy in the European parliament from England. He claimed that support for minority languages was illegal under European law and the support was withdrawn. It is our hope that when Europe takes a formal vote on this issue in the near future this support will be restored. [What was it that Europe paid for?] It paid fifty percent of the cost of transportation to the school for the children who needed it and also transportation costs for inter-village visits. Now the families have to pay for that. The poorest families, however, cannot afford to pay for this service. This is unfortunate because we are convinced that it is important for the children in our schools to meet children from schools in the south in their home context. In the light of this we are opening a new school in La Guingueta [an important border town known as Bourg-Madame in French] next year. I believe this branch will be a real European school since it will take children from both sides of the border. [Are any of the parents who send their children here Catalan activists? Do some of them support local autonomy for Northern Catalunya?] Of the eighty-six children in this school only two sets of parents are politically involved. But I think that politics are changing in France. I believe that our success is not only a sign of the rebirth of Catalan culture in the north but also a symptom of a deep crisis in France. I do not believe that France can survive in its present form under the evolving structure of the European Community, because it is the only country in Europe that is so highly centralized. [What do the children do, as far as the Catalan language is concerned, when they reach eleven years of age and leave for junior high school and high school?] In this town, the authorities have begun what they claim is a bilingual program in the secondary schools. In our opinion it's not really bilingual but is better than nothing. However, our children will be bored in programs of this type because they know so much Catalan. Now we have been told that the children who come from bilingual schools and from *Les Bressoles* will be put in a different, more advanced class. We hope this will be the case. [How are the teachers prepared? What degrees do they have?] We have two methods of preparation and that's another reason we prefer to remain private and independent. On our own we have created a normal school in Béziers that specializes in regional languages (Occitan, Catalan, Basque, Alsatian) as well as the usual preparation for teaching in elementary schools. This institution is private, but when its graduates teach in our schools they are paid the same wages and carry the same titles as those teachers who have the state diploma. The school is our UFM. [Special university for the education of elementary and high-school teachers.] This school has been functioning since 1996. In our school we now have three teachers who are graduates of this center. We also have some teachers who were trained in the *Principat*. Unfortunately, their degrees do not carry the same weight as those obtained in France and therefore, in most instances, they suffer lower wages than our French trained teachers and, additionally, cannot be considered for tenure. They have the status of temporary teachers. We protect them as best we can, however, because we don't want to lose them. While we pay them the same salaries as our regular teachers, they do not receive any funds for retirement. Thus, in real terms, they earn about one quarter less than those with French diplomas. In general there are no conflicts between them and the other teachers because they understand the situation. We would of course prefer to correct this inequality and have talked with the minister of education about it, but up to now it has not been possible to make the change. This is very unfortunate because I really feel that, apart from language, the pedagogy involved in training teachers in the *Principat* is better than in the French system. The teaching examination in France is based on skills in the French language and math. There is almost no pedagogy involved. Graduates of the French school start to learn pedagogy when they are already teaching after graduation. In the *Principat* they are trained to be teachers first and then, only afterwards, take their examinations. This is one of the reasons for the problems in French education right now. In our own Teachers' College we have added one year of obligatory teaching before our students take the examination. To achieve this we signed an agreement with the University of Girona [in the *Principat*] that allows our students to study there for one year and then enter our own Teachers' College. Unfortunately, we don't have any students in this program right now because we don't have any money to support the living expenses of students during their stay in Girona. I talked with the Catalan government about finding some support for them, perhaps through such employment as the teaching of French. The

Generalitat has expressed an interest in helping out. [Do you teach other languages in the *Bressoles*?] When the children reach the age of eight we give them six hours of French per week. In this respect it is worth noting that our students do better in French on the average than the students who have gone to French public schools. I think the reason for this difference is that after a student has mastered the speaking and reading of Catalan it's then easier to learn to read and understand French.

[What is your opinion of the UCE?] I think that it's very important. [What does it do for those interested in Catalan culture?] The idea behind the UCE was thought up by Raymond Maillol now president of the *Bressoles*, but there is no connection between the UCE and us. The UCE functions as a link between the different Catalan-speaking regions in the south and, to a lesser degree, with Northern Catalunya. In the beginning when things for Catalan were so difficult, it was important for the people in Prades to see so many people speaking Catalan; thinking in Catalan in all subjects. It was important to show people in Prades that it's possible to speak Catalan here. About ten years ago there was a conflict between the local people and the UCE, but now they want the UCE to stay. When the directors of the UCE were discussing the possibility of changing the location of the school, the people of Prades went to the mayor and told him to do what he could to keep the UCE there. There are, of course, economic reasons for this. More than one thousand people are present for ten days during summer to take courses at the UCE. This represents an important economic gain for Prades. I also think the people of Prades appreciate the behavior of students at the UCE. The UCE is a nonviolent way of opening doors.

[How would you describe Catalan identity?] It's hard to say. We are nonviolent and, given this fact, the peoples of the world have to respect us for surviving so many aggressions throughout our history. We have lost many times against those who were out to destroy us. The French and Spanish defeated us in 1714. I hope in the new Europe we will gain respect from all the inhabitants of the continent. Europe is *the* hope for all of us.

Interviewee 5. Woman, retired elementary schoolteacher who volunteers in her community to teach Catalan to adults. Interviewed in French in Céret in 2000.

[Where are you from?] I was born on the other side of the frontier near Prats de Molló. I came here when I was very young, but Catalan is my maternal language. I learned French in school and later took Spanish because I could not take Catalan at that time. [Can you give us some history behind the adult education program in Catalan here?] The present school was founded by Mr. —— when there was no formal program in Catalan. Before his retirement he was a professor of Spanish in a local school. As early as the 1950s his department was allowed to offer students other Hispanic languages, notably Portuguese and Catalan. I started to teach Catalan in the school system after passing the required oral examination in 1964. There were about fifteen of us who took the test. The professors who taught Catalan and graded the examinations at that time had degrees from Paris, Toulouse, or Montpellier. I took my examination in Montpellier. It was called a "Certificate in Catalan" and allowed those of us who passed it to teach the language in junior or senior high school. When the head of the Spanish department retired in 1981, a group of people in our town asked him whether he would be willing to teach Catalan to adults. He agreed to serve without pay and the city council provided a space for his course free of charge. Soon afterwards he also organized a Catalan theater group, again on a volunteer basis. He soon realized that he had taken on more than he could chew and sought out an assistant to help with the language course. I agreed to join the program. [Who are the people who take Catalan here?] It's a bit of everything. In the beginning it was mostly non-Catalans who had moved here from elsewhere, but that has recently changed. When I began to teach, most of my students were retired people from other regions of France. They heard Catalan spoken in the stores and in the weekly market and were sorry that they couldn't understand it. When you don't understand a language you don't know if others are speaking badly of you. For the past several years Catalan has been declining and is now spoken primarily on market days when you will also hear several foreign languages spoken by many of those retired people who come from beyond the borders of France. The Vallespir [a more remote part of Northern Catalunya] is now the region where Catalan is spoken the most. In the Cerdanya, Catalan is also quite often the language spoken by the natives. They don't even realize that they are speaking it, it's so natural for them. In any case little by little people from this town began to join the Catalan classes. Many of these were functionaries, former professionals, who grew up here but had migrated to work in some other part of France. Many of them had been elementary schoolteachers during the time when Catalan was forbidden in their classrooms. Other people who take the course already speak Catalan but do not know how to read or write it. For example, I now teach an advanced class for people who are fluent in the spoken language. One interesting thing about this class is that the students often discover they know local expressions that they learned as children but that are not in the standard language. They often find these discoveries quite surprising. [How many students does your school have from year to year in all the classes?] Last year we had sixty-four students on four levels. There are now three of us who teach these different levels, all without pay, of course. Our school is supported by the municipal office of culture that grants us funds by way of the town council. This allows us to buy necessary supplies for our courses such as textbooks. We also have a free subscription to a children's review that is sent to us by the *Generalitat*. Additionally we have set up a lending library. The mayor here is Catalan and many of the councilors are as well. They support us willingly but *not* too enthusiastically. There are no conflicts over Catalan but culture in general is not very well supported in the region. [Is tourism important here and what influence does it have on Catalan culture?] Tourists come looking for something new and different. It's not tourists who are the major problem. The real problem is people from the outside who have moved permanently from other regions and have no interest in local culture. Many of these are retired folks who first came on vacation and then bought permanent residences. This has had a major negative effect on the strength of Catalan culture in the region. [How would you define being Catalan here?] To be Catalan you have to feel Catalan. It's not a question of being from here or from elsewhere. There are locals who don't feel Catalan at all. It's mostly the elderly who feel the strongest ties to Catalan culture. Most of the youngsters see themselves as French and nothing more. When I ask young people if they feel Catalan they say no. In fact it is most likely the case that if Catalan is still spoken it's because many people from the south have settled here. [Do you have any contacts with the south?] No, not very many but the local Sardana group does have, and it's a very large association. It's only recently, however, that the Sardana has become popular. It came to us via the south. The current Sardana group has strong ties with various groups in the *Principat*. There are also Catalan lace makers who come to our market from the south to sell their wares. I might add that during village festivals, exchanges with the south occur frequently. In order to increase contact with villages south of the border our language classes take trips to the *Principat*. [Are there any contacts between your association and other Catalan teaching associations?] Yes, there are. Most of the schools that give adult courses in Catalan are part of a collective group. All of us prepare students for the international certificate in Catalan. There is also an *Omnium Cultural* in Perpignan that depends on the *Omnium Cultural* in Barcelona although it is an independent association. The Perpignan association sends people to teach Catalan in towns that do not have any of their own teachers. In such cases the students have to pay to take the courses. [Do you know how many villages and towns have classes in Catalan?] No I don't know because we are independent and municipal. We existed before the *Omnium* was created. We are not the only independent school. Many parents are now requesting bilingual schools for their children and some of them have asked to have adult classes so they can practice Catalan with their children. There is one bilingual public school each in Perpignan, Prades, and Céret but it's a very recent development. These bilingual programs begin in pre- and elementary schools up to the sixth grade. The teaching is half in French and half in Catalan. Here in preschool, the children have alternate mornings in Catalan and French. There are not enough places to fill the demand, however, because these programs are not well funded. Parents don't necessarily see these schools specifically as a means of learning Catalan but rather so their children can be exposed to a foreign language at an early age. These classes only continue up to the age of eleven. After that age, in the junior high school, Catalan is no longer taught for the same number of hours as French. There is generally one course specifically offered in Catalan and, additionally, one other subject taught in Catalan. Here one of the courses is usually history because the history teacher also speaks Catalan. In the high school, students can choose an option in Catalan and apply this to their diploma if they pass the final examination in all subjects. At present, in our adult courses, we also have a few foreigners. One is German and the other is English. We use an existing English method for teaching Catalan for our English students. We do not have enough teachers, however. [Did you always speak Catalan when you were young?] Yes, I frequently visited my grandparents in the south during vacations and always spoke Catalan with them because they did not speak French at all. I still keep up contacts with other relatives in the south. Additionally, here at home, I speak Catalan with my husband. But when we married we lived outside the Catalan-speaking area and for that reason I did not speak it with my children. Now they regret that. My younger sister practically never spoke Catalan. [Her husband comes in, joins the conversation and says that when he and his siblings lived together with their grandparents in a big house, they all spoke Catalan. He continued;] We spoke Catalan at table. But later in our own house we spoke French with our parents. My brother speaks Catalan poorly and my sister does not speak it at all. Catalan is no longer spoken in my parents' village which, over time, has become a suburb of Perpignan. [We return to the teacher.] [Can you tell us a bit more about the parents who send their children to Catalan schools? Do they ever speak up about autonomy or independence for the region?] Oh no, no, no! Of course not, only the children of Catalans go to these schools. But there is no relation whatsoever between politics and these schools. Catalans here are not like the Corsicans or the Basques. In general Catalans are more moderate than nationalists from other regions. Perhaps that's why we are lagging behind in bilingualism here if you compare us with other places that have their own languages. [Have you been to the UCE?] Yes, my parents live close to Prades so it's easy for me to attend courses there. [What do you think is its importance and/or its function?] It was created in the region to have an effect on the local mentality. I am not sure that this has happened, however. But the school is very interesting. Among other things you can meet people there from everywhere Catalan is spoken. As far as Prades is concerned the UCE aids the local economy. Some of the people in Prades, however, are afraid that the UCE attracts people who have a fixed idea concerning Catalunya. There is no real Catalan community here. [Her husband then interjects.] It's not something we think about because we are all Catalans. Our son speaks Catalan but he does not use it at all. [She then rejoins the conversation.] Students in our classes often attempt to speak Catalan with business people here but get discouraged because they are often criticized for their efforts. Some even say to them that it's Spanish they are learning and not Catalan, or that they speak it very badly! But those who really want to practice their Catalan go to the market in Figueres, across the border. There, no one criticizes them but tend to respond to them in French, because they have French accents. When our classes go on excursions we speak Catalan among ourselves. If we invite a guest speaker, for example, to the local library, we ask them to speak in Catalan if they can. But if there are people in the audience who do not speak Catalan they usually ask that the talk be given in French and so it usually is. [Are there people here who are ashamed of having Catalan roots?] Now no, but before, yes, particularly during the exodus from the Spanish Civil War. Now it's the reverse. People who never wanted to speak Catalan in the past now want to speak it. It has become the language of the intelligentsia.

It is obvious that the above interviewees are not from the mainstream of Northern Catalunya's current inhabitants. They—along with their friends and colleagues of the same "persuasion"—do, however, represent the minority of individuals committed to the survival of Catalan culture that does exist to maintain and promote their cultural identity, in spite of the overwhelming pressure against it. Some despair, but nonetheless remain combative. One of our interviewees, for example, in spite of his cynicism, is a vigorous supporter and administrator of the UCE and remains committed to the struggle.

Chapter Eight

Language and Identity on the Horns of a Dilemma

In this, the final chapter, I shall examine some of the problems raised by demands on the part of *Catalanistes*, for independence or, at the very least, a greater degree of autonomy, including stronger protections for the Catalan language in all spheres of public life. I shall also consider the periodic dangers faced by Catalans who see their identity as something other than Spanish in the context of a frequently hostile, centralist nation-state. To illustrate these problems I have given voice to four individuals all of whom are Catalan speakers, each with a unique point of view of these issues.

The first interview is with a well-known fiction writer and Catalan militant, a member of the ERC, Maria Mercè Roca, who won a seat in the Catalan parliament as a deputy from Girona in the elections of 2003. In the interview she expresses her strong fear that unless the laws concerning language use in Catalunya are strengthened, it is in imminent danger of giving way to Castilian. She favors independence for Catalunya.

The second interview is with a man born and raised outside of Catalunya, a native speaker of Castilian, who now speaks fluent Catalan with great pleasure. Although not opposed to the status of Catalan as the co-official language in the *Principat* he is against the imposition of the language in public life and in the school system. His is an interesting case because, unlike many Castilians who take up residence in Catalunya but who feel no need to learn the language, he enjoys speaking it with Catalan friends and never imposes his Castilian on them.

The third interview is with Tilbert Stegmann, born in Barcelona during Franco's reign to German parents, who has become a strong advocate for the Catalan language and the separation of Catalunya from Spain. He is a professor of linguistics in Germany and militates for the language in both of his native countries.

The last interview is with a strong Catalanist, a priest and native of València who teaches religion in the Valencian public school system.

He presents his views of the language problem in his territory and discusses how it is linked to local and national politics.

The Interviews

Interviewee 1. Maria Mercè Roca, born in Portbou. Both her parents are from Catalunya. Roca left home at the age of seventeen, going to Girona where was a resident student in a Catholic school. She now resides permanently in Girona where she practices her profession as a writer. Recorded in Girona in September 2003.

[Why do you write in Catalan?] For sentimental reasons. I feel Catalan and I am Catalan. My social life has always been based on the language, although all of my education, which took place during the Franco period, was in Castilian. [Can you write in both languages?] Not really. For me Castilian is a second language. I can write letters in Castilian, but make many errors in it. I don't really feel able to write it properly, at least for literary texts. Catalan is the only language in which I can express my literary ideas. I am not really bilingual. [What is your opinion of bilingualism in Catalunya?] I believe that bilingualism is a very bad solution. When two languages are spoken in the same space one of them always takes precedence over the other. In the actual situation today Castilian is taking precedence over Catalan in many areas of culture. Under Franco, Castilian was the official language and many activities were carried out in it but Catalan remained very much alive. Now, a change is taking place even though Catalan is co-official and is described as our "native language." For example, the language of the justice system is essentially Castilian. All the judges and lawyers use it. The dominant language in shops and restaurants is also Castilian. There is a demographic problem here. Catalunya has always been open to immigrants and accepting of everyone. For this reason (which is not a bad thing) it creates dangerous conditions in which we don't have a basis to protect what is ours with laws so that people from the outside will have to accept our rights and responsibilities. The government has allowed bilingualism to exist in order to win votes. Even in areas that are "reserved" for Catalan, Castilian is creeping in. Take, for example, TV 3 which in principle is a Catalan-speaking station. Now one hears more and more programs in Castilian, and of course there are many other stations available to the public that are primarily in Castilian. TV 3 should be exclusively in Catalan.

This is both a political and a psychological problem linked to the loss of prestige. Many young people who speak Castilian do not regard Catalan as a "modern" language. In reality Catalan is the language of literature, culture and, in principle, of the university. On the other hand, the language of the street, of work, and of leisure is Castilian. Also there are very few films in Catalan. When a native Castilian speaker who understands Catalan is present in a group of Catalans most people will revert to Castilian anyway. They renounce speaking their own language (Catalan). Only a minority of people engage in bilingual conversations. This is a very serious problem. Those who do not wish to abandon their own language find this a very difficult thing to stick to. Even people who are strongly *Catalanistes*, some of whom would vote for independence, immediately switch to Castilian when they show up at work. [What is the reason behind this?] It's a problem of insecurity. Also, for so many years in the past people were convinced that Catalan was only a little thing. Many Castilian speakers feel that Catalunya is just a part of Spain. They don't care if Spain does not defend the rights of Catalunya. If the central government really cared about, or was obliged to respect, plurilingualism it might work but they do not support these ideals.

[What about Catalan intellectuals?] Many of them have good values and ideas but these have to be translated into action. They also are too often ready to make compromises. [And you as an intellectual?] I don't see myself as an intellectual. I see myself as an artist. With language I invent worlds. In another country I would merely be a writer! But here, by the simple fact of having to live as a writer who lives in a world of two languages, and writes in Catalan, I am less honored by the public than those who write in Castilian. It costs an author who writes exclusively in Catalan a great deal because the major means of communication are Castilian. [What do you think of Catalan writers who write in Castilian?] There is, of course a question of liberty, but it pains me nonetheless. Novels are written in one language or another but a language is also a political fact and has political weight.

[Why have you decided to enter politics?] Look, I would be much more tranquil staying home and writing. However, there are things that one cannot abandon. The Catalans need help to protect their culture. Catalan is a language that does not have a state with the power to stand behind it. Society here is becoming more and more Castilian. It affects me body and soul. [For you, the language is in peril?] Yes, yes, yes. Recently *El Punt* published an article clearly showing that in just a little time the language has diminished. There are many people who come here from the outside and they assimilate very little into Catalan culture. [Don't their children learn Catalan in school?] Of course, but outside of school they speak Castilian with one another.

Additionally, there is a problem in higher education. Many universities do not have enough students for the number of courses they offer. For example, Catalan philology is given in many places but few students take it. There is also the fact that a university professor can chose to teach in either Catalan or Castilian. Usually in a class of say thirty students, if only one of them asks for Castilian, the course will be given in Castilian. Some professors do refuse to cede to the lone student and insist on teaching in Catalan, but the majority of them accept the change.

[Girona seems to us to be a city that is very Catalan. Do you agree with our judgment?] Yes, that is the case. In Girona even the street is very Catalan. Barcelona, on the other hand, is totally different. There, many people are aggressive about being Spanish. [What about waiters and workers in shops here?] The situation is mixed but here there are many who do speak Catalan.

[What do you think about the new coach of the Barcelona football team and his language program?] Under the new coach the team members will be obliged to understand Catalan. I believe that this is very important since in every country of the world those foreigners who want to earn their living must understand and speak the local language. Also, the new rules for the team provide a method for foreigners who join the team to integrate into Catalan culture. And it's not a difficult thing to impose. Membership alone on the team is also a method of integration. Another recent positive development is the growth of the ERC in Catalunya. The governing party up to now, the CiU, has never pushed enough to enforce its Catalanist programs. One thing that is still lacking here is financial help via foundations supported by industrialists who could provide funds for the struggle. After all, they do make their money here and they could return some of it by contributing to Catalan culture.

[Are there other writers who are interested in supporting *Catalanism*?] Not many participate in the public sphere but there are a few who serve as advisors in some aspects of government. Also a few writers publish critical articles in the newspapers. Every day these journalists complain about the current situation. They help readers to maintain their self-esteem in the face of the hostility and negative attitudes of so many Castilians who live amongst us.

Interviewee 2. Retired former functionary who spent a good deal of his professional life working in the *Principat* as a representative of the national government. Now retired he continues to live in Catalunya. Though he speaks excellent Catalan, as well as several other languages, he remains a Spanish nationalist. Recorded in Girona in September 2003.

[How did you learn to speak Catalan?] It was easy. I love languages. I learned it talking to people. For many years I lived with a Catalan family and the language of the household was exclusively Catalan. At work I had to speak Castilian since most of my colleagues were from other parts of Spain. On the other hand, after work I mixed with the local people and, conversing in Catalan with them, the language came quickly. Now I always speak Catalan with my Catalan friends I still feel Spanish even though I am not a nationalist. I love to learn foreign languages and if I had a chance to learn Chinese I would do it. [What do you think of Catalans?] They are too closed in and too nationalist. I believe that the idea of independence for Catalunya is a stupid one because it will cost too much. It's really a cliché. What will happen if they have their independence! It's not possible. And why bother, with a Spanish passport you can go anywhere. Besides, Catalunya gets a great deal of financial help from the central government. [But is it not true that Catalans give much more in taxes to Spain than what is returned to them and don't in any case receive as much aid back from Madrid as other regions?] It's possible, but I don't really know about that. I think that the autonomies in Spain have lots of privileges. Girona, for example, has the highest standard of living in all of Spain. The people in the Balearic Islands have a very good quality of life. I think that these complaints are exaggerated. [Do you have any friends who are *Catalanistes*?] Ah, no I really don't. Yes there is one. He really does feel "Català Català." I suppose that he wants a Catalan passport and is ashamed to carry a Spanish one. He is not a person who likes to dialogue with others about his beliefs. [Do you feel that the Catalan language can survive?] Of course! The way Catalans defend it I see no reason to doubt that it will continue to be spoken. It will always be preserved in the home by the people. After all Franco did forbid its use in public but it continued anyway. [Yes, but when there is an enemy it's easy to preserve things of one's own culture is it not?] That is true, but here it is the official language of the school system. This causes an unfair situation for Castilians families living here. For example, when those from the outside have to put their children in school, the children will lose out because they do not speak the language. And this is not at all a rare occasion. Lots of people come to Catalunya from elsewhere, functionaries for example. I believe that classes should be in Castilian for everyone all over Spain and then children who want to can take the local language as an elective. There are people who come here from Andalusia and do not know a word of Catalan. Why should their children have to take math in Catalan when they should be taught in Castilian. The same goes for geography or history or any other course outside of language courses.

Languages should not be imposed on people, but people should learn as many as they can.

Interviewee 3. Tilbert Stegmann, professor of philology at the J. W. Goethe University in Frankfort. He was born in Barcelona during the Franco regime. He is the author of the book, *Catalunya Vista per un Alemany*. Recorded at the UCE in 2002.

[Why are you so interested in Catalan culture and language?] There is a biological reason. I was born in Barcelona during the Franco period. My father was director of the German High School there. Throughout my childhood I was a victim of Franco's propaganda and, as a result, neither Catalan culture nor its language existed for me. No one ever spoke to me in Catalan when I was young. It was only many years later when I went to the university that I came to realize that there was another language in Catalunya different from the one I knew (Castilian). It was not until I was in my late twenties that I began to learn Catalan. I was working on my doctorate in Hamburg and had to go to Catalunya to do research for my thesis. Then back in Hamburg, I discovered a young co-student who did not speak Catalan even though she and her family had also lived in Catalunya. So we decided to help each other to recover "our" Catalan language and soon began to use it exclusively when we conversed. My other motivation for learning the language was the shock I felt in finding out that a dictator could prohibit a language in such a way that it had been hidden from me during those ten years that I had lived in Barcelona. I also learned that Germans in general had no idea of the cultural difference between Catalunya and the rest of Spain and, as a result, I found myself somewhat obliged to become the "ambassador" of Catalan culture in Germany.

I had many opportunities to acquaint Germans with Catalan culture. For example, in 1978, I convinced the Berlin Senate to donate money that was put aside for international summer programs dealing with culture to fund a three-week program presenting Catalunya to the public by way of a festival of Catalan culture. By some miracle a quarter of a million marks were given to support this effort and, additionally, nearly the same amount was given by Catalunya. The Berlin senate also made available the entire infrastructure of the Berlin Festspiele to organize the entire program that consisted of one hundred different activities. Even today this festival that generated a great deal of interest in Catalan culture is legendary. Never again has such a presentation of Catalan culture taken place anywhere else in the world outside of Catalunya. All the first class writers came to participate as well as all the great political leaders, linguists, etc. It was a great opportunity for me, a chance to get to know all these people. Joan Fuster [major Valencian figure of *Catalanitat*], for example, who was known to never leave his village in València, came when the Catalan minister of culture, Max Cahner, promised to accompany him personally.

The major thing about the festival was that it was keyed to the general German public rather than to a small group of scholars. [What was the reaction of the Catalan authors who participated?] I remember Joan Brossa [famous Catalan visual artist] who was a great admirer of Wagner, coming to the opera with me. Montserrat Roig [well-known writer of novels in Catalan and author of an important book on Catalan prisoners in Germany during the Hitler period] gave a public lecture in the Library of Berlin on the subject of the extermination of Catalans in the Nazi camps. It was a very emotional experience. It is also worth noting that, at least partially as a result of the festival, there are now thirty-five universities in Germany that have programs in Catalan. This fits in, however, with the linguistic interest in Romance languages, as a whole, that has traditionally exited in Germany.

[Are you optimistic or pessimistic about the future of Catalan culture and language?] I am optimistic by nature. As long as Catalan culture exists I will continue to work for it. But to do this, one has to have an objective view of the existing difficulties. There are many at present. With Aznar [president of Spain at the time of the interview] things have gotten much worse. Jordi Pujol [president of the *Generalitat* at the time of the interview] sent me a copy of a recent lecture he gave. The topic was "Good and bad things about politics." He noted in it that politicians are negatively viewed by the majority of people who don't appreciate the auto-censure that, they have to employ. But if they say what they really think the few remaining doors open to them will close. Pujol made it clear that when he communicates with President Aznar he cannot fully express his opinions.

[Are you aware that when elementary schoolchildren in Catalunya meet in the schoolyard for recreation they revert to speaking in Castilian?] Yes, this often occurs, although they all speak Catalan well. Typically, when they leave the school context they converse in Castilian. This is a very serious problem. We have to make Catalan more "chic" for the young. The truth is that while Catalan is necessary to succeed in the professions it's not seen, for example, as the language of discotheques or other leisure activities.

[What do you think of the UCE?]. It's wonderful! Valencians come to recharge their connection to Catalan culture and language. It has an important function for the "in-group" and is a necessary reality. It supports the participants' pride in Catalan culture.

[What is your opinion of Catalunya in the context of Europe? Is it positive or negative?] There are different perspectives. The most realistic one is that, given the fact that the hands of Catalans are tied by the Spanish Constitution, they have little leeway in which to pursue their goals in Europe. The optimists see Europe as a hope but if the right is in power, there will be very little chance to do more than waste a great deal of energy to gain as much as possible under the circumstances. If the Socialists win the next election, I don't know if by then they will have come to understand the lessons of the past demonstrating that more liberty for Catalans is good for all Spaniards. Juridically, the central government has the right to stop the move toward greater liberty but morally they are obliged to respect it. Unfortunately Catalunya has to get permission of the state to make changes in its statutes. Such changes will first have to pass a referendum in Catalunya but even in this case, if it is declared unconstitutional by the Spanish supreme court, the demand can be refused. The majority of Spaniards are centralistic. My realistic optimism tells me that the Catalans will have to put a great deal of pressure on the central government to make it accept the Catalan point of view. If the ERC were to double the number of votes let us say to twenty percent of the electorate, the Catalan bourgeois or nationalist parties, with the help of the ERC, might be able to accomplish something.

[Do you believe that in the context of Europe Catalunya can play an important role?] Patience will be necessary in the coming years during which the flame of Catalunya must continue to burn among the Catalan majority or Catalunya is dead. But there is very little enthusiasm for this among the people. Most feel that Catalunya is in an excellent position with its education, with its parliament, its urban centers, its TV, although the latter, for example, should be much more Catalan than it really is. Josep Lluís Carod-Rovira [president of the ERC] a friend of mine, is very clever politically, but his party needs to win more seats. They have to convince ordinary Catalans that their accommodation to the current situation is not helpful enough for progress. Catalans have to fight more for their identity and their cultural personality and they must realize that, if they do this, nothing bad will happen to them. To the contrary, they still fear that if they act, the Spanish Army will be sent after them.

[Are the people really afraid?] It's not that they are afraid—rather they continue to cling to their old habits that inhibit action. They don't want to complicate their lives. Catalans are well-educated, good business people, and know how to communicate with everyone. This is a positive factor that gives them certain advantages, but also disadvantages. I fear that they are a bit too accomodating. They should be much less so in those areas that would improve their situation. If Catalans spoke only Catalan in public, for example, it would help the situation since it would provide more consciousness of the fact that they are different from the rest of Spain. But I fear that it will only be a small minority of the elite that will, perhaps, slowly bring the Catalans forward in the context of Europe, hopefully leading to a peaceful separation from Spain. Catalans have made great progress in teaching immigrants to speak the Catalan language but this does not mean that the latter use it in everyday practice. The UCE serves to create the elite that I am speaking about but it remains too small. Pasqual Maragall [now president of Catalunya] is something of a *Catalanista* but he worries about Madrid and the opinion of the Spanish Socialist Party [the PSOE]. If the Catalans do not stick together they will fall. The left in Catalunya is still very fragmented. If they could get together they could emphasize the fact that their relation to Spain is a *contract* rather than a marriage.

[Do you feel that independence is possible for Catalunya?] Yes, with all that I have said, Catalunya can achieve independence just like Estonia and the other Baltic States. But the legal and constitutional situation makes it very difficult. It is quite a heavy task and needs immense patience as well as perseverance to say to the Spaniards: We are well brought up and convinced democrats, but we are not Spaniards. One has got to become accustomed to hearing this every day! It's a question of democratic rights.

Interviewee 4. A priest who teaches religion in Catalan in the Valencian public school system. Recorded at the UCE in 2002.

[Is Catalan taught in the public schools in València?] I teach my courses in Catalan, although I do not teach it as a language. My courses are on religion and specifically Catholicism. I have some colleagues who also teach in the language. In principle there are other courses in the language that involve total immersion. [At this point he begins to call the language Valencian but, as we shall see below, he uses the word alternately with the word "Catalan" to describe it.] While Valencian is my first language I had to learn to write it as an adult because it was officially forbidden by Franco. [Did you speak Catalan in your house?] Of course, it was the language of the entire village. It was not until the 1980s that the language was recognized as co-official by the Valencian government. Now in large cities the language is being lost due to a great deal of immigration. It is also deprecated as the language of peasants and, therefore, is often looked down upon, particularly in urban areas. However, in spite of its repression the language did survive the Franco regime. [Are courses in Valencian obligatory in the schools?] The law says that by the time they graduate from high school, children are supposed to be fluent in both Valencian and Castilian. However, although there is considerable demand for the language, it is often not offered because of a lack of teachers. Many teachers in the public school system are from outside of València and do not know the language. Now that the PP is in power in València the government does nothing at all to improve this situation. The law is clear but it is not respected. [Don't people denounce this lack of respect for the law?] Yes, they do, but the government does not seem to care. Even when parents request it, it's frequently impossible to find someone to teach in the language. Actually there is more demand in València for Valencian than for Castilian. [We have no way of verifying the last statement.] To say otherwise is not telling the truth.

[What are the parties in order of size in València?] First is the PP, then PSOE, then ERC, then the nationalist block. But it is very difficult for small parties to gain representation. All parties are required to win at least five percent of the vote throughout València to be represented in the parliament. This effectively keeps them from ever being elected.

[In the Church what language is used?] There are, of course native speakers of Valencian who are priests, but there is a shortage of clergy in general and the authorities never assign us priests who are Valencian speakers. They come from other parts of Spain and the mass is always said in Castilian, even in communities in which Catalan [note the change in naming the language] is the common tongue. There is a great deal of self-hatred in regard to Valencian culture. This is a major problem. It's very sad. There are many people who are natives who have abandoned the language of their parents.

[Who are the immigrants who come to València?] In the past most of them were from other parts of Spain, now many are from South America. They are Spanish speakers and Catholics. I work for an association that teaches Valencian to immigrants. But the ones who speak Castilian don't have any interest. Madrid favors immigrants from South America over those who come from non-Castilian-speaking countries. I believe that Madrid sends many of these people to us as a means of diluting our language and culture. [He then goes on to explain why he believes that this practice is designed to defeat Catalanism in all of the Catalan-speaking autonomies.] València is long but very narrow. It has a long frontier with Castile. Thus, there is a great deal of contact between the two areas. The Catalan territories are like a frying pan. You can put lots of potatoes into the round part, but the handle is very narrow. So the central government attacks València because it is the handle which is so much more vulnerable. Ever since Philip the Fifth the central government has been attacking València to weaken Catalan culture in its most vulnerable place. If València falls then the *Principat* will fall a bit later.

The current authorities in València have never recognized the integrity of the Catalan language. There is a very large internal problem in València because so many people fail to realize what is going on. When Franco died and the new constitution was passed for Spain the Spanish nationalists put a clause in the constitution that said neighboring states within Spain could not sign agreements to federate. [We have verified this in the Spanish constitution.] About the same time people began to say that Tarragona, to the north of València, is Catalan-speaking but just to the south, across the border in València, the language is Valencian. [Are there conflicts between the north and the south?] They have recently begun to appear but in the past this was not the case. Negative feelings here about Catalans to the north are fostered by clichés concerning their attitudes and comportment. [Do you talk about all this in your classes?] Yes, I certainly do talk about this and I put it in a historical context.

[Do you see the UCE as an important institution?] Yes, it's very important. Outside, propaganda against the UCE often refers to it as a band of terrorists, but the UCE occupies an important place in the fostering of Catalan culture. [If someone asks you what your identity is, what is your response?] Catalan! And I have a Catalan identity card which, of course, has no legal status.

Self Hatred in Catalan Culture

A complaint expressed during our interviews by those who consider themselves to be *Catalanistes* is that Catalan culture, in general, is hurt by the feeling of self-hatred or insecurity expressed in different ways by so many Catalans. This problem is discussed at length in the book *Autoestima i Països Catalans* by the following four authors: Quim Gibert, a psychologist, Josep Murgades i Barceló, a philologist, Bernat Joan i Marí, professor of linguistics and literature, and Marc-Antoni Adell, an inspector of education. As Bernat Joan i Marí says,

> A Catalan speaker carries a "computer chip" programmed in his collective memory of three centuries of the direct repression of his language and the millions of times there's been a negative reaction to his use of

Catalan. It's possible that he will prefer to abandon the use of the language—or not use it every time when it is convenient for him to do so—or even come to hate those who use it or finally hate all Catalan speakers (even himself). (98)

The author then goes on to list a set of mechanisms that contribute to linguistic inferiority. Among these are the following:

1. The language is the language of peasants rather than the educated.
2. The language sounds ugly to the ear.
3. To speak a language in certain circumstances is a sign of bad manners (a factor particularly common among Catalans when in the presence of Castilian speakers even those who speak or, at least, understand Catalan).

Generalizing from these principles at the end of the book, Bernart Joan i Marí points out that self-hatred is a powerful tool for the maintenance of an established power relation within states that have linguistic communities with minority status. These ideas support the conclusion that language is a solid aspect of cultural identity and to attack it is to attack the culture in general. He also notes that a language can serve as a strong integrative tool in the integration of foreign immigrants into the local cultural scene. This is, of course, precisely why the Catalan government has made language the major criterion of Catalan identity and has fostered programs to teach Catalan to those foreigners who wish to integrate.

Albert Branchadell and the Case Against Independence for Catalunya

No one can accuse Albert Branchadell of anti-Catalan sentiments. As noted in chapter five, Branchadell is a sociolinguist, and one of the organizers of an association that lobbies for the inclusion of Spain's three regional languages on four important official state documents: driver's licenses, postage stamps, identity cards, and passports. Yet, in his book, *La Hipòtesi de la Independència*, he argues against independence for Catalunya on two grounds: one pragmatic; the other, moral. The pragmatic reason cited is that independence would have the perverse effect of putting the Catalan language in serious danger. He notes that this is not the case, under autonomy (the present status quo) because Catalan is protected by the European Community's regulations (accepted by the central Spanish government) concerning recognized national (regional) languages. Independence, on the other hand, he warns, would create a situation in which Castilian speakers living in Catalunya could demand the creation of separate schools to teach all subjects in Castilian. Such an outcome would clearly be a defeat for both Catalan culture and the integration of Spanish speakers into that culture from other regions of Spain as well as from foreign Spanish-speaking countries. Although it is my opinion that his pragmatic argument carries great weight and must be considered seriously by those Catalans in favor of independence, I find his *moral* argument against independence (to be discussed below) seriously lacking in historical depth, considering the number of times Catalan culture has been put in jeopardy by Madrid, most recently in the 1990s when Aznar and his party took over the Spanish parliament with an absolute majority. (In his previous four-year term, Aznar had required the support of the CiU and Basque nationalist representatives and was, therefore, forced to tread lightly on nationalist feelings.) Even today, in the Catalan-speaking areas of València and the Balearic Islands, there is a concerted effort on the part of the ruling party, again the PP, to subvert Catalan culture and language. For example, in València and in the Balearic Islands the local public television stations have become mouthpieces for the ruling party and have overtly attacked the status of Catalan in their territories. In València the government and its language "experts" have consistently pushed the claim that València is a language separate from Catalan in spite of opinions to the contrary by the vast majority of professional linguists and literary figures. Most recently, *El Temps*, carried an article in their January 3, 2005 issue stating that the Valencian government had created a major crisis in the Valencian Linguistic Academy when it declared that it would take legal action against the academy if it were to declare that Catalan and Valencian were the same language.

Branchadell begins his demonstration that independence for Catalunya would be immoral by arguing that the Catalan language is in no danger in the *Principat*. To support this claim he cites a series of statistics taken from a range of recent and past questionnaires on the use and linguistic capacities of citizens within the territory that show a constant statistical improvement in the knowledge of Catalan in the population. Though I have a great deal of confidence in Branchadell, I know him to be a careful scholar and have benefited personally from the two courses I took with him at the UCE, our approach is quite different. I must confess that as an anthropologist it is difficult for me to trust the results of most questionnaires. In general, I prefer to use long and partially open-ended interviews conducted only after I have developed a social relationship with respondents. The shotgun approach of questionnaires leaves them open to doubt because respondents often give answers they believe to be in conformity with the wishes of the administrators of such protocols. I must add, as well, that in our experience of living for three months in Barcelona and of visiting other large cities in the *Principat*, such as Terrassa, Tarragona, and Lleida, we heard very little Catalan spoken in commercial spaces, in the streets, and in public conveyances. We were, it is true, on the other hand, often impressed by the frequency of Catalan in the smaller cities of the interior, such as Manresa, Berga, Vic, and Olot. The most recent study concerning the use of Catalan as the language of the home, carried out in 2004, came to a more negative conclusion. (These results are also drawn from a questionnaire as subject to doubt as all the others and which I cite for what its worth.) What this study purports to show that only 51 percent of the respondents habitually used Catalan in conversations with the members of their families!

What ever truth lies in the statistics concerning the frequency of the language in Catalunya I am most startled by Branchadell's argument that independence for Catalunya would be *immoral* when he cites the fact that other countries have managed to recognize and accommodate to two and in some cases three languages within their territories. This is, of course, true of Finland that recognizes Swedish in the Aland Islands, even though its inhabitants represent only about 10 percent of the national population. Belgium, admits co-equal status for Flemish (Dutch), French, and in (a small) eastern part of the country, German. Switzerland has four official languages: German, French, Italian, and Romansch, the ancient Romance language of the country spoken by a tiny minority of the population. Canada recognizes French and English as official languages although French speakers are almost exclusively limited to the provinces of Quebec and New Brunswick. The reader should note, however, that all of these languages with the exception of Romansh, are spoken as the official languages of other nation-states and, in the context of the European Community, are accepted as official languages of that body. Neither Basque, Catalan, nor Galician, defined within Spain as co-official languages with Castilian, are likely to ever be accepted by the European Community as official languages of that body. Any vote on the rules concerning this change would require an absolute majority of all members. Even if the Spanish government were to back such a request, its adoption is an unlikely event because it would surely be blocked by France.

To bolster the moral side of his argument Branchadell cites the Canadian political scientist Stephane Dion who states the following:

a) Secessionist movements are rooted in two types of perceptions: the *fear* inspired by the union and the *confidence* inspired by succession.

b) Secessions are improbable in well established democracies because these two perceptions are unlikely to exist simultaneously at a high level of intensity. (73 in Branchadell, italics in original, and the text is quoted in English)

Branchadell feels that neither of the conditions cited by Dion is met by the current situation in Catalunya. This point is important in the sense that it is unlikely that Catalunya will, in fact, achieve independence in the near future, or indeed if it ever will. But, in my opinion this is beside the point. The movements in Catalunya that militate for independence do so in the context of an autonomy that adheres to democratic principles and which consistently bases the notion of national identity on language and the freely chosen desire of inhabitants to be members of such a new state. Catalan nationalists have frequently called for referendums on the issue of independence in order to determine the will of the people who live and are entitled to vote there. Where is the immorality in this?

There is no evidence that Catalan nationalism is in any way based on racism or any other kind of genetic argument for the attainment of citizenship status. Therefore, it seems to me that no existing Catalan political party or cultural group should by Branchadell's arguments abandon this desire for separation from Spain no matter how utopian such a position might be at the present time. The real question (in the light of Spanish history, ancient and current) is: Can the Spanish government give permanent and stable assurances, written into its constitution, that those regions of the nation-state recognized after the death of Franco as historical autonomies, and granted special status within the Spanish state, will never again be threatened with the loss of this status? And will future Spanish governments be ready to consider that certain new rights, democratically won, will be seriously considered in the context of what is essentially a federal system?

Appendix

El Cas De Catalunya
(The case for Catalunya)

The following text is taken from an English version of the petition to the United Nations meeting in San Francisco in the month of April 1945 in which the U.S. delegation of the *Consell National Català* presented the case for granting independence to Catalunya under the sponsorship of the newly formed United Nations.

To the United States of America, The United Kingdom of Great Britain and Northern Ireland, The Union of Soviet Socialist Republics and the Republic of China, Sponsors of the United Nations Conference on International Organization, at San Francisco.

In as much as Catalonia (in spite of her present subjugation under Spain) is a well defined nation, as proven by her history, her ethnological characteristics, her particular language, her own literature and culture, her specific laws, her customs and traditions and, above all, her permanent and manifested will and her desire to regain national sovereignty;

In as much as Catalonia (because of her unrecognized status as a nation) cannot adhere to the United Nations Declaration nor declare war on any Axis power and so gain admission to the San Francisco Conference;

In as much as Catalonia, being occupied by the Fascist armies of General Franco, cannot proclaim her *de facto* state of belligerency against the Nazis nor gain official recognition for her many sons fighting now in the "United Nations" Armies;

In as much as Catalonia cannot, in justice, be classified as a neutral nor legally as an Ally, although she is a friendly nation still occupied by Nazifascism;

In as much as, on the other hand, the legal institutions of Catalonia no longer exist her President Lluís Companys having been executed by Franco and her democratic government disbanded and nullified;

We, therefore, in our own name as members of the Catalan National Council (United States Delegation) in the name of 75,000 organized Catalans residing on the American Hemisphere and in the name of the people of Catalonia whose voice is now expressed

REQUEST from the sponsors of the San Francisco Conference:

That in view of these special circumstances and the unique position of Catalonia; in view that Catalonia is one of the few remaining nations in Europe whose national rights have not yet been recognized, CATALONIA BE CONSIDERED A SPECIAL CASE and since she cannot be legally represented nor actually participate in the proceedings of the Conference, BE ALLOWED TO REPRESENT AND FILE THE FOLLOWING APPEAL TO THE UNITED NATIONS before their representatives at San Francisco.

APPEAL TO THE UNITED NATIONS ON BEHALF OF CATALONIA

The special situation of Catalonia as a nation prevents her from having legal representatives and from being present at this Conference to participate with the United Nations in the charting of an international organization for peace and security. But it is precisely in view of Catalonia's unique situation that we have decided to present her case to your attention so that the national rights of 3,000,000 Catalans may be known to all the United Nations and may be justly considered in the charter for a new world. We are not presenting a problem of frontiers or political reconstruction, economical recovery or any such matters which are not to be attended to until after the security situation has been set up. We present a case for national liberty which requires solution or at least consideration precisely while the security negotiations are taking place.

Catalonia existed as a free nation until 1714, when she was finally incorporated into the Spanish State and is one of the few remaining nations in Europe whose national rights have not yet been recognized. This makes the case of Catalonia almost an obsolete problem, mainly because most of the problems of the European nationalities were supposedly solved at Versailles. But Catalonia was not, her liberties were not restored at the end of World War I like those of the other nationalities. It is for this reason that we present here her problem as a special case and as a matter which demands consideration and solution so that the Charter which is to be written for the new Europe does not become once more an injustice against Catalonia.

In a way, Catalonia's over-prolonged captivity and retarded liberation is due, more than to several military defeats, to repeated diplomatic misfortunes suffered by Catalonia. In 1713, by the Treaty of Utrecht, after a long war against her Spanish oppressor, Catalonia's rights were disowned by her own allies and sacrificed to power politics and matters of expediency. In 1919, at Versailles, in spite of the 18,000 volunteers Catalonia contributed to the Allied armies, the rights of that unfortunate nation were once more overlooked. In 1924, at Geneva, due to the defective clauses of the League's Covenant, and to the presence of Spain in the League of Nations, which made impossible the required vote of unanimity on such matters, the League of Nations could not even consider Catalonia's demands for liberty. In 1937, at the Nyon Conference, Catalonia's rights were disregarded but, on the other hand, the claim of Italy about "the right to intervene in Spain to prevent the setting up of an independent Catalan Republic" was considered valid.

This traditional diplomatic indifference towards Catalonia's claims should not lead anyone to believe that the Catalan case has no bearing in the maintenance of permanent peace and security in Europe. The "Catalan Question" has been at the bottom of much of the unrest and political turmoil in the Iberian Peninsula during the last three centuries, and there has never been any aggressive power or force in Europe which has not tried, at some moment, to speculate on the Catalan discontent to enhance or secure success of their plans. As an example, we will mention that during the period of the French Revolution, Robespierre in person, with the aim of gaining Catalonia to his cause, visited Barcelona with the written "Constitution of Catalonia" in his brief-case. Scarcely a quarter of a century later, Napoleon Bonaparte, in order to gain a foothold on the Peninsula, actually created a "Catalan State" and tried to establish a Catalan government separate from that of the kingdom of Spain. In recent times, in the geopolitics of Germany for the Mediterranean area, Greater Catalonia (that is to say the old Catalan Kingdom or the present territories of Catalan language—Catalonia, València, French Catalonia, and the Balearic Islands) was scheduled to play a big role against France and her African empire, although Catalonia did not accept the "New Order" or Nazism. As a final proof of the importance of Catalonia in the stability of Europe, any well-informed and clear-sighted statesman will admit that lest the national problem of Catalonia is satisfactorily solved, there will never be real peace and order in the Iberian Peninsula.

Now that a charter of the nations of the world is going to be definitively written for a durable peace, Catalonia cannot let this opportunity pass without appealing to the justice of the United Nations for her due recognition, lest new and irreparable mistakes are committed at the moment of the charting, and her national freedom be postponed indefinitely.

In appealing to you for justice, Catalonia wants to state her full aspirations and the full scope of her rights. We earnestly request that this Conference not commit the same mistake of the Peace Conference at Versailles, which dismissed Catalonia's demands on the erroneous argument that it was a mere case of home rule, to be granted by Spain, and as such an internal problem, a "family conflict," to be solved within the Spanish State. Neither could we allow the United Nations to judge Catalonia's rights on the same basis of the League of Nations that classified them as a simple problem of a minority inside Spain. Catalonia is a nation and must be recognized as such before any regional organization can be established in Spain, in the Iberian Peninsula, the continent of Europe, or the liberated world.

It is unnecessary to tire your attention by detailing the historical, ethnical, linguistic, and cultural reasons which prove the national characteristics of Catalonia; nor need we present any list of all the persecutions and oppressions of which she is and has been a victim. Neither do we believe it necessary to adduce proof of her determination and ever-increasing will to live again as a free nation. We do not even need to point out that her struggle for freedom has continued through the centuries. In 1640, in the first attempt against her freedom, Catalonia fought against Spain (War of Succession) and proclaimed the Catalan Republic; in 1714, after her national rights were all disregarded at Utrecht, Catalonia continued to fight against Spain and France even after having been abandoned by her allies (England, Austria, Portugal, and Holland); in 1931, Catalonia led in the Peninsula the democratic and civil revolt against the Bourbon Monarchy, proclaimed a Catalan Republic and made possible the Spanish Republic; in 1931, Catalonia organized a national plebiscite in which 98 percent of the population proclaimed Catalonia's will of self-government; in 1934, after the Spanish Republic had fallen into the hands of the fascists and reactionary forces, Catalonia struck for democracy and national freedom and proclaimed the Catalan State as part of the Confederacy of Iberian Nations; in 1936, upon the Nazifascist coup of Franco and the Falange, Catalonia became the bulwark of antifascism and fought at the same time for her national freedom.

There is, however, a matter which should be set forth here very clearly to end all possible misconceptions. We refer to the permanent and inalterable nature of the Catalan problem. In other words, the basic terms of the Catalan aspirations do not change with the existence of a more or less liberal regime in Spain, nor with a greater or lesser degree of persecution or oppression. For instance, Catalonia's aspirations are independent of the existence or non-existence of Franco in Spain. Catalonia has been an oppressed nationality under the Monarchy, under the Spanish Republic, and under Franco. The removal of Franco *alone* will not solve the Catalan national problem, as it was not solved by the overthrowing alone of the Bourbon Monarchy. Catalonia fights Franco and it's trying to overthrow his fascist regime, and in the same spirit she is fighting for the destruction of Hitler and Hirohito. Franco is Catalonia's present tyrant, but in the dual function of representative of Nazifascism and representative of the Spanish Unitarian State. The replacement of Franco will free Catalonia from Nazifascism but it will not make Catalonia free from Spanish oppression.

This takes us to the real danger of any Catalan "solution" carried out under the light of routine misconception. For this reason we appeal herewith to the United Nations before any decision be taken and any commitment be made with regard to Spain. Too many people erroneously believe that the Catalan case is merely a Spanish problem. It is not so. There is a tendency to classify the Catalan question among the internal problems of Spain. The conflict between Catalonia and Spain, as any problem between any oppressed nationality and her oppressor, has always been of an international nature. Catalonia is not a Spanish conflict but a European problem. In these terms the problem of other European nationalities were solved at the peace table at Versailles.

To class the Catalan question among the internal problems of the Spanish state is to appoint Catalonia's oppressor to be sole judge in a conflict in which it is itself a contending part. Catalonia, or any other oppressed nation, cannot expect justice from her own oppressor.

As history shows, not even a liberal and democratic Spain, of the type of the Spanish Republic, is capable of solving the Catalan national problem. Most of the leaders and the statesmen of the former Spanish Republic live in the erroneous notion that the principles and the clauses of the Atlantic Charter apply to the Spanish State but do not apply to Catalonia and the other nationalities incorporated by force into the Spanish State.

Hence Catalonia cannot accept the premise that her national liberty is to be identified and confused with the problems of restoring democracy and restoring the republican regime in Spain. At its due time the United Nations will have to confront the Spanish problem, and Catalonia will help with all her strength to solve it, but its denomination and solution has no direct relationship with Catalonia's problem of national liberty.

Let no one misinterpret this statement. Catalonia is vitally interested in democratic Spain. 150,000 Catalan youths died in the Spanish Civil War to eradicate Fascism from Spain and to secure the subsistence of democracy in the Iberian Peninsula. But it is as Catalans that the people of Catalonia want to participate in the welfare of the Iberian block of peoples. They want their rights as a nation to be recognized so that Catalonia, through self-determination, can be free to join in the political reorganization of the Peninsula. Once free and duly recognized as a nation, Catalonia will be in a position to consider, for instance, a Confederacy of Iberian States, on the basis of equal rights and voluntary association, in which Catalans, Basques, Galicians, Spaniards and Portuguese could participate.

On the other hand, Catalonia being absolutely identified with the cause of the United Nations—in whose armies so many of her sons are fighting on all the fronts—declares her willingness to accept the sacrifices the reorganization of Europe may demand of her, no matter how jealous she may be of her sovereignty and freedom as a nation.

Summing up, Catalonia REQUESTS from the United Nations:

a) THAT her delayed case of national liberation be, from this moment, scheduled as one needing immediate solution.
b) That her case for self-government be filed for immediate solution under the principles and the clauses of the Atlantic Charter, independently of any regional solution contemplated for Spain.
c) THAT her position in the political organization of Spain be decided by herself, through plebiscite of the Catalan nationals, after recognition of her status as a nation.
d) THAT any further disagreement or dispute between Catalonia and Spain be submitted for hearing before the United Nations Council or the Internal Court of Justice on its behalf.

In submitting her claim for national liberation before this Conference and before the international public opinion, Catalonia expects justice from the United Nations. (New York, April 14, 1945)

To this declaration I wish to add the footnote that the address given on the document presented to the U.N. is: *Consell National Català*, 239 West 14th Street, New York, NY. This is now the address of *La Nationale*, a Spanish nationalist organization!

Bibliography

Abley, M. 2003. *Spoken Here: Travels Among Threatened Languages*. Boston and New York: Houghton Mifflin Co.
Anderson, B. 1991. *Imagined Communities*. London and New York: New Left Press.
Barton, S. 1993. The Roots of the National Question in Spain. *The National Question in Europe in Historical Context*. Edited by Makuláš Teich and Roy Porter. Cambridge, UK: Cambridge University Press.
Bauraoui, N. 2000. *Garçon Manqué* [Meant To Be a Boy]. Paris: Stock.
Ben [the artist] à Céret . 1997. Interview published by the Museum of Modern Art, Céret [France] in conjunction with his exhibition.
Branchadell, A. 2001. *La Hipòtesi de la Independència [Hypothesis of Independence]*. Barcelona: Empúries.
Casanova, E. 2003. *Viatge a les Entranyes de la Llengua: de la Riquesa a l'agonia del Català* [Voyage into the Inner Reaches of the Language: From its Richness to its Agony] Lleida (Catalunya): Pagès Editors.
Castellvi, J.N.D. "Les Foguerades Patriòtiques": La Invenció d'un Nou Ritual Identitari (1906–1907) [The Flaring Up of Patriotism: The Invention of a New Ritual of Identity]. *Música i Cultura Popular: Caramella*. 68–75.
Connor, W. 1994. *Ethnonationalism: The Quest For Understanding*. Princeton, NJ: Princeton University Press.
Cardús i Ros, S. 1997. The Role of Nations in the European Construction. Translated by Cardús i Ros.
Derrida, J. 1996. Monolinguism of the Other or the Prosthesis of Origin. Translated by Patric Mensah. Stanford, CA: Stanford University Press.
Di Giacomo, S.M. 1987. La Caseta i l'Hortet [The Shed and the Vegetable Garden]. *Anthropological Quarterly*. 60, no 4. 160–166.
Gelner, E. 1983. *Nations and Nationalism*. Ithaca, NY: Cornell University Press.
Gibert, Q., J. Murgades, M.-A. Adell, and B. Joan. 2003. *Autoestima i Països Catalans* [Self-Esteem and the Catalans]. Barcelona: La Busca Edicions.
Graves, L. 1999. *A Woman Unknown: Voices From a Spanish Life*. Great Britain: Virago Press.
Hamilton, H. 2003. *The Speckled People*. London and New York: Fourth Estate.
Hobera, J.R. Catalan National Identity: The Dialectics of Past and Present. *Critique of Anthropology*. 10, no. 2, 3. 11–28.
Joan i Mari, B. 2003. Enfocament Sociolingüístic. *Autoestima i Països Catalans*. Barcelona: La Busca Edicions. 89–107.
Lamming, G. 1983. *In The Castle of My Skin*. Ann Arbor, MI: Michigan University Press.

Lebovics, H. 1992. *True France*. New York: Cornell University Press.
Lluís, J.-L. 2002. *Conversa amb el Meu Gos* [Conversation With My Dog] Barcelona: La Magrana.
McRoberts, K. 2001. *Catalonia: Nation Building Without a State*. Oxford, UK: Oxford University Press.
Mosse, G. 1995. Nationalism and Racism. *Nations and Nationalism*. 1, part 1. 163–74.
Noyes, D. 2003. *Fire in the Plaça: Catalan Festival Politics After Franco*. Philadelphia: University of Pennsylvania Press.
Pla, J. 1997. *How the French Established Their Identity in* Cerbère. *Contraband i Altres Narracions*. Barcelona: Edicions La Butxaca, 111.
Roca, M.M. 1991. *La Casa Gran [The Big House]*. Barcelona: Columna.
Shakespeare, W. *The Tempest*. 1964. Edited by Robert Langbaum. New York: Signet Classics, Penguin Books.
Stegmann, Til. 1986 (new edition 1996). *Catalunya Vista Per un Alemany* [Catalunya As Seen By a German]. Barcelona: La Campana.
Strubell i Trueta, M. 1984. Language and Identity in Catalonia. *International Journal of Society and Language*. 87. 91–104.
Terradas, I. 1990. Catalan Identities. *Critique of Anthropology*. 10, no. 4. 39–50.
Tree, M. 2000. *CAT. Un Anglès Viatja Per Catalunya per Veure si Existeix* [An Englishman Travels Through Catalunya to See If It Exists]. Barcelona: Columna de Viatge.

Supplimental Bibliography

The following list consists of a selection from the books consulted as background reading for this work.

Books in English and French on Nationalism and Catalan History

Bessiere, B. 1992. *La Culture Espagnole, Les Mutations de l'après— Franquisme (1975-1992)* [Spanish Culture. Changes During the Post-Franco Period]. Paris: L'Harmattan.
Conversi, D. 2000. *The Basques, the Catalans and Spain*. London: Hurst and Company.
Generalitat de Catalunya. 1995. *Language Rights and Cultural Rights in the Regions of Europe: Proceedings of International Symposium*. Barcelona: Generalitat de Catalunya, Department of Culture.
———. 1996. *Everything About Catalonia*. Barcelona: Biblioteca de Catalunya.
Hobsbawm, E.J. 1990. *Nations and Nationalism Since 1780*. Cambridge, UK: Cambridge University Press.

Hughes, R. 1993. *Barcelona*. New York: Vintage Books.
Lafont, R. 1967. *La Révolution Régionaliste* [Regional Revolution]. Paris: Gallimard.
Le Roy Ladurie, E. 2001. *Histoire de France des Regions* [The History of French Regions]. Paris: Editions du Seuil.
Orwell, G. 1952. *Homage to Catalonia*. Boston: The Beacon Press.
Sahlins, P. 1989. *Boundaries, The Making of France and Spain in the Pyrenees*. Berkeley, CA: University of California Press.
Woolard, K.A. 1976. *Double Talk: Bilingualism and the Politics of Ethnicity in Catalonia*. Stanford, CA: Stanford University Press.

Books in Catalan and French dealing with Political and Linguistic Issues in Catalunya and between Catalunya, Spain, and France

Ainaud de Lasarte, J.M. 1995. *El Llibre Negre de Catalunya: de Filip V a l'ABC* [An extreme right-wing newspaper]. Barcelona: Edicions La Campana. Negative citations by Spaniards and the Spanish press from 1714 to 1994. (All citations in Castilian.)
Alemany, J. (translator) 1999. *Forum Babel, El Nacionalisme i les Llengües de Catalunya* [Babel Forum. Nationalism and the Languages of Catalunya]. Texts written by Antonio Santamaría in Castilian. (The *Forum Babel* was a pro-Spanish group in Catalunya claiming that Castilian was in danger in the *Principat*.)
Alexandre, V. 1999. *Jo No Sóc Espanyol* [I am not Spanish]. Barcelona: Proa.
Anguera, P. et al. 1996. *El Catalanisme Conservador* [Conservative Catalanism]. Girona: Quaderns del Cercle.
Balcells, A., E. Pujol, and J. Sabater, 1996. *La Mancomunitat de Catalunya i l' Autonomia* [The Mancomunitat of Catalunya and Autonomy]. Barcelona: Institut D'Estudis Catalans.
Boyer, H. 1991. *Langues en Conflit: Études Sociolingüístiques* [Languages in Conflict: Sociolinguistic Studies]. Paris: L'Harmattan.
Cabana, F. 1996. *La Burgesia Catalana* [The Catalan Middle-Class]. Barcelona: Proa.
———, ed. 1998. *Catalunya i Espanya. Una Relació Econòmica i Fiscal a Revisar* [Catalunya and Spain. An Economic and Fiscal Relationship must be Revised].
Cabana et al. 1999. *Balears, Catalans, Valencians* [The Catalans, the Valencians, and the People of the Baleares]. Barcelona: Portic.
Candel, F. and J.M. Cuenca, 2001. *Els Altres Catalans del Segle XXI* [The Other Catalans of the Twenty-First Century]. Barcelona: Planeta. Candel, a Catalan of Castilian origin, and author of *Els Altres Catalans* [The Other

Catalans] (1964) and *Els Altres Catalans Vint Anys Després* [The Other Catalans Twenty Years Later](1985), looks to the future of Spanish immigrant assimilation into Catalan culture.

Duran, M. 2001. *França Contre els Catalans: L'Afer Bonnet* [France Against the Catalans: The Bonnet Affair]. Barcelona: Llibres de l'Índex.

Ferrer i Gironès, F. 2000. *Catalanofóbia. El Pensament Anticatalà a Traves de la História* [Catalan Phobia. Anti-Catalan Thinking Throughout History]. Barcelona: Editions 62.

———. 2005. *El Gran Llibre per la Independència* [The Grand Book for Independence]. Barcelona: Columna.

Granell, F. et al. 2002. *Catalunya Dins la Unió Europea* [Catalunya Inside the European Union]. Barcelona: Editions 62.

Gubert i Macias, J. 1990. *Portbou, Segle XIX*. Porbou: Ajuntament de Portbou.

Huguet, J. 1999. *Cornuts i Pagar el Beure. El Discurs, Anticatalà a la Premsa Espanyola* [Cuckold and Paying for the Drinks. Anticatalan Discourse in the Press]. Barcelona: Columna.

Junqueras, O. 1998. *Els Catalans i Cuba* [The Catalans and Cuba]. Barcelona: Proa.

Molas, I. 2001. *Les Arrels Teòriques de les Esquerres Catalanes* [The Theoretic Roots of Catalan Parties of the Left]. Barcelona: Editions 62.

Rafanell, M.A. 1999. *La Llengua Silenciada* [The Silenced Language]. Barcelona: Empuries.

Riera, I. 1998. *Els Catalans de Franco* [The Catalans of Franco]. Barcelona: Plaza i Janés Editores.

Roca, F. 2000. *Theories de Catalunya: Guia de la societat Catalana Contemporania* [Theories of Catalunya: Guidebook to Contemporary Catalan Society]. Barcelona: Portic.

Roig, M. 1977. *Els Catalans als Camps Nazis* [The Catalans in Nazi Concentration Camps]. Barcelona: Editions 62.

Ruiz, F., R. Sanz, and J. Solé i Camardons. 1996. *Historia Social i Política de la Llengua Catalana* [Social and Political History of the Catalan Language]. València: Perez Climent.

Simó, I.-C. 2000. *Carta al Meu Nét: Sobre el Nacionalisme* [A Letter to my Grandson About Nationalism]. Barcelona: Columna. (Simó, of Valencian origin, writes to her fictive grandson to tell him why she is a Catalan nationalist.)

Simon i Tarres. 2005. *Construccions Polítiques i Identitats Nacionals* [Political Constructions and National Identities]. Barcelona: Biblioteca Abat Oliba. Publicacions de l'Abadia de Montserrat.

Strubell i Trueta,T. 2000. *Josep Roca i Ferreras i l'Origen del Nationalisme d'Esquerres [Josep Roca i Ferreras and the Origin of Leftist Nationalism]*. L'Hospitaler de Llobregat: Arenys de Mar.

Termes, J. 1999. *Les Arrels Populars del Catalanisme* [The Popular Roots of Catalanism]. Barcelona: Empúries.

Vallverdú, F. 1998. *Velles i Noves Qüestions Sociolingüístiques* [Old and New Sociolinguistic Questions]. Barcelona: Editions 62.

Vila Casas, E. and P. Candel, 1996. *Memòries d'un Burgès i d'un Proletari* [Memories of a Bourgeois and of a Proletarian]. Barcelona: Columna. (The book consists of two life histories about being and becoming Catalan during and after the Franco period. Vila Casas writes in Catalan, and, in this book, Candel writes in Castilian.)

Vilar, P. 1977. *La Catalogne dans L'Espagne Moderne* [Catalunya in Modern Spain]. Paris: Flammarion.

Poetry and Fiction

For a general introduction to contemporary Catalan fiction, I recommend to those who can read Catalan:

Bordons, Glòria and J. Subirana. *Literatura Catalana Contemporània* (Barcelona: Proa, 1999).

Espriu, S. 1978. *Primera Història d'Esther Antígona*. Barcelona: Editions 62. (Originally published during the height of the Franco period in 1948.) A very difficult long poem written by one of Catalunya's greatest modern poets. The History of Esther retells the biblical story of Esther but is, more importantly, a disguised cry against the Franco Government and the repression of the Catalan language.

Two Classics Translated into English

Martorell, J. and M. J. de Galba. 1984. *Tirant lo Blanc*. Translated by David H. Rosenthal. Baltimore: Johns Hopkins University Press. An epic of the Crusades published in the sixteenth century, this classic story won high praise from none other than Cervantes who recognized it at the time as "The best book in the world."

Rodoreda, M. 1980. *La Plaça del Diamant*. Translated as *The Time of the Doves* by David Rosenthal (1986). Saint Paul, MN: Graywolf Press. This classic novel set in Catalunya at the time of the Spanish Civil War has been translated into thirty-seven languages.

Other Fiction that Deserves Translation

Roig, M. 1987. *La Veu Melodiosa* [The Melodius Voice]. Barcelona: Editions 62. A short surrealistic masterpiece set in the Franco period concerning a young man brought up in his grandfather's house and home schooled by a series of brilliant teachers to keep him from being contaminated by Franco's lies. Finally leaving home he encounters the real world with mixed results.

Sales, J. 1982. *Incerta Glòria*. Barcelona: Editions 62. A classic novel about the Civil War in Spain.

Index

Abley, M., 5
Adell, Marc-Antoni, 185
Algeria, 1, 2, 66, 87, 149
Allègre, Claude, 144
Alsace, 4
Amics de Catalunya, 73–74, 126–27
Anderson, Benedict, 38–40, 51, 55, 63
anthropology, 7–10
  changing theories, 7–8
  postmodernism and, 8–9
  relativism and, 7, 9
architecture, 16, 22, 24, 35, 46, 60, 98, 116, 118, 138, 155
Argelès, 14, 72, 89, 135–36
Arrels, 68, 152, 165
assimilation, 38, 41, 64, 139, 146, 158, 160, 177
Autoestima i Països Catalans, 185–86
Avui (newspaper), 18, 44, 50, 57, 117, 131, 148, 158
Aznar, José María, 49, 145, 148, 181, 187

Balearic Islands, 18, 48, 110, 123, 126, 129–30, 166, 179, 187, 193
Banyuls
  architecture, 116
  Catalan language and, 79, 80, 82
  Cerbère and, 24, 67, 73, 76–77, 83, 88, 89, 91
  housing, 90
  Portbou and, 97, 103, 112
  sports, 29, 69
  tourism, 15
  wine production, 15, 79
Barcelona
  architecture, 46
  choral singing and, 60–61
  Corts of, 45
  democracy and, 47–48
  football, 61, 75, 99, 107, 178
  language study in, 17–19, 59, 153, 162
  Olympic Games in, 49
  Portbou and, 13, 21, 24, 33, 82, 91, 115
  Robespierre and, 193
  speaking of Catalan and, 4, 73, 77, 83, 84, 131
  TGV train line and, 23
Barton, Simon, 41–43
Basques, 5, 18, 46, 47, 49, 98, 123, 133, 155, 188, 196
  Aznar and, 187
  language, 74, 157, 168
  media's view of, 125–26
  Spanish Civil War and, 58
  suppression of, 55
Baylac, Alà, 122, 140–41, 142
Bec, Pierre, 148
Benjamin, Walter, 22, 103
  See also Walter Benjamin Foundation; Walter Benjamin monument
bilingualism
  Alsace and, 4
  Catalunya radio and, 19
  children and, 26, 80, 94, 96, 102
  confusion caused by, 23
  education and, 4, 152, 167, 172
  families and, 114, 120, 172
  Portbou and, 20, 26, 94, 98, 102, 105, 112
  public opinion of, 176–77
Blanc, Jacques, 124, 148
Blanc, Paul, 124
Bonaparte, Napoleon, 4, 193
Bonet, Maria del Mar, 61
bourgeoisie, 39, 55, 182

Bouraoui, Nina, 1
Branchadell, Albert, 121–22, 125
   and case against independence for Catalunya, 186–89
Bressoles, 152, 164–66, 168–69
Bricker, Jordi, 117
Brossa, Joan, 181
Bru, Lluís, 60

Cahner, Max, 124, 181
Canal 33, 18, 48
Canigó, 31–32, 98, 110, 121, 138, 164
Cardús i Ros, Salvador, 50–51
Carod-Rovira, Josep-Lluís, 43, 182
Casals, Pau, 120, 134
   Museum, 138
Casanova, Eugeni, 131, 144–46
Castellers, 59–60
Castilian, 1, 12, 18–19
   Catalans and, 41, 43, 94–99, 100–6, 108–13, 153
   cultural dominance of, 176–78, 186
   nova cançó and, 61
   as official language, 45, 47, 48–49, 83, 98, 125
   in Portbou, 20, 22–23, 26, 29, 33, 94, 98
   schools and, 66, 70, 82, 94, 95, 99, 130–31, 165, 181, 184, 187
Catalan-Aragon alliance, 44–45
Catalan history
   end of Franco's reign, 47–49
   important dates in, 44–47
Catalan identity
   Canigó and, 110
   in Cerbère, 26
   choral singing and, 60
   Europe and, 169
   language and, 53, 91, 99, 186
   nationalism and, 51, 55–56
   in Portbou, 33
   Ruc and, 62
   Sardana and, 28, 59
   self-hatred, 138–39, 185–86
   study of, 11–12, 14
"Catalan Identities" (Terradas), 53–55
Catalan language
   French attitude toward teaching of, 142–44
   study of, 144–46
"Catalan National Identity, The Dialects of Past and Present" (Hobera), 55–56
Catalan nationalism, 12, 35, 49–51, 62–63, 81, 91, 108, 120, 161, 189
   rebirth of, 51–57
   symbolic forms of, 57–62: Barcelona football, 61; *Castellers*, 59–60; CAT Internet designation, 62; choral singing, 60–61; excursionism, 62; national holidays, 60; *nova cançó*, 61; *Sardana and the Cobla*, 58–59; Savings Bank of Catalunya, 61; *Segadors*, 57–58; symbolism, 62
Catalanism. *See* Catalan nationalism
Catalanistes, 124, 175, 177, 179, 185
Catalonia, Nation Building Without a State (McRoberts), 45
Catalunya-Ciutat, 52
Catholic Church, 47, 118
   Franco and, 150
   language and, 45, 184
Cerbère
   associations in, 67, 68, 78, 80, 89–90
   as case of lost culture, 90–92
   commerce, 27–28
   demographics, 25–26
   German occupation, 71, 74–75, 86
   overview, 14–16
   Portbou and, 13–14, 25–35: symbolic links, 29–30;

festival of St. Jean, 30–32;
  proposed tunnel, 24, 25, 91
social life, 28
sports, 28–29
Céret, 81, 83, 136–38, 152, 172
Charter for Regional or Minority
  Languages, 143–44
Chevènement, Jean-Pierre, 135, 158
children, naming of, 140–42
Chirac, Jacques, 125, 136, 147, 154
choral singing, 60–61, 83
church and state, separation of, 140
Civil Guard, 102, 107, 109–10, 112
Cobla, 58
Colera, 20, 24–25, 34, 75, 91, 100, 103, 107, 110
Companys, Lluís, 47, 191
Connor, Walker, 40, 55
Convergència Democràtica de Catalunya, 47
Convergència i Unió (CiU), 33, 49, 122, 178, 187
Corsica, 45, 142, 144, 151, 155, 158, 173
Council of Europe, 143–44
Creoles, 39
cultural relativism, 9
customs bureau
  Cerbère and, 68, 71–72, 80, 91
  closing of, 13, 115
  Portbou and, 94, 99, 101, 104, 111

Dadaism, 8
Dalí, Salvador, 20, 106, 137
Darwinism, 8, 54
Declaration of the Rights of Persons Belonging to National or Ethnic, Religious, or Linguistic Minorities, 142–43
  Article 1, 142–43
  Article 2, 143
  Article 4, 143
Derrida, Jacques, 2
Diada, 60
DiGiacomo, Susan M., 51–53

Dion, Stephane, 189
Domènech i Montaner, Lluís, 46, 60
Druon, Maurice, 4–5

education
  cooperative school project, 81, 82
  school exchange program, 86, 87
El Punt (newspaper), 18, 157, 177
El Temps (magazine), 18, 83, 117, 142
environmental issues, 70, 88, 123
Espriu, Salvador, 61
Esquerra Republicana de Catalunya (ERC), 118, 123, 175, 178, 182, 184
ethnography, 8–9
ethnonationalism, 40, 55
European Community (EC)
  autonomous regions and, 108
  Cerbère and, 35
  financial support, 164, 167
  language law, 121, 125, 187, 188
  license plates and, 118
  and naming of children, 140
  open border policy, 13, 130
  Portbou and, 24, 34
  Prades and, 124
  regional languages and, 154
European Language Charter, 154
excursionism, 62

Fabra, Pompeu, 46, 120
Ferdinand (king), 42, 45
Ferreres, Itgnasi, 1
festivals
  Catalan boat festival, 95
  Catalan cultural festival, 180–81
  Cerbère and, 30, 34, 68, 74, 78, 84, 90
  Festival Major, 95
  music, 120, 138
  Northern Catalan and, 161, 171
  Patum, 127–28
  Portbou and, 96–97, 114
  Sardana and, 28

festivals—*continued*
  See also Sant Joan festival; *Sant Jordi* festival
football, 23, 28–29, 61, 69, 75, 98–99, 107, 178
  See also *penya*
France-Deux, 4
France
  attitude toward teaching Catalan, 142–44
  cultural identity, 146–47
  educational system, 157
  regional languages and, 147–48
  Vichy government, 136, 147
Franco, Francisco
  Abbey of Montserrat and, 19
  Casals and, 120
  Catalan and, 4, 12, 56, 73, 99, 102, 106, 113, 118, 176, 179
  Church and, 150
  economy and, 16
  ERC and, 53
  events following death, 33, 47–49, 53–54, 128, 158
  exiles and, 135, 138, 160
  Llach and, 81
  Majorcan and, 3
  nationalism and, 43, 50
  nova cançó and, 61, 161
  petition to U.N. requesting independence for Catalunya, 191, 194–95
  Portbou and, 104
  red list and, 71
  Sardana and, 58–59
  smuggling during reign, 12, 14
  Spanish Civil War and, 30, 47, 58, 72, 75, 101
  St. Jean festival and, 31
  Valencian and, 183–84, 185
Frèche, George, 147–48
Free Zone, 78, 79
French Revolution, 133, 140, 150, 193
Fuster, Joan, 181

Gaudí, Antoni, 46
Gelner, Ernest, 37–38
Generalitat, 24–25, 33–34, 43, 47, 54, 77, 101, 169, 171, 181
  Portbou and, 91, 115
  school funding and, 152
  support for Catalunya, 56–57
  UCE and, 123–24, 163–64
Gibert, Quim, 185
Giralt, Emili, 124
Goethe, J.W., 2, 180
Gras, Felix, 147
Graves, L., 3
Guifré the Hairy, 44–45
gypsies, 100, 134, 145–46, 158

Hamilton, H., 3
Hobera, Josep, 55–56
housing, 34, 116, 124
  in Cerbère, 67, 68, 70, 79–81, 85
  in Portbou, 105, 109, 114

Ibarretxe, 125–26
Ibsen, Henrik, 46
identity cards, 104, 118–19, 154, 185, 186
Imagined Communities (Anderson), 38
Infopista, 49
Iniciativa per Catalunya Verds-Esquerra Unida (IC-V-E-U), 123
Institute of Catalan Studies, 46, 162–63
INTEREG, 164
Internet, Catalan language and, 49, 62, 153
Invasion of the Barbarians, 7
Isabel (queen), 42, 45

Jacobins, 116, 149, 159
Jaume I (king), 45
Joan i Mari, Bernat, 185–86
Jocs Florals, 46, 51, 63
Juan Carlos (king), 47

Judaism, 11, 75

Karavan, Dani, 22, 98
See also Walter Benjamin monument

La Jonquera, 101, 107
labor unions, 17, 74, 88, 129, 166, 189
See also railroad
Ladino, 2
Lamming, G., 2
"Language and Identity in Catalunya" (Strubell i Trueta), 57
language laws, 48–49, 101
European Community and, 121, 125, 164
opinions on, 109, 113
Languedoc-Roussillon, 85, 122, 123–24, 147, 151–52
Larzac, 12, 17, 127,
La Gran Mentida (Carod-Rovira), 43–44
Lebovics, Herman, 146–47
Les Caixes, 61
Les Voltes, 117–18
Liber Ludiciorum, 44
Liber Maiolichinus, 44
Llach, Lluís, 61, 81
Lluís, Lluís, 5
Louis II (king), 45

Macià, Francesc, 52–53
Maillol, Raymond, 169
Mancomunitat de Catalunya, 47, 63
Maó, 129–31
Maragall, Pasqual, 125–26, 183
March, Ausias, 61
Marquez, Gabriel Garcia, 10
Matamala, Feliu, 118–19
McRoberts, Kenneth, 45
Mitterand, François, 135, 166
Moore, Michael, 132
Morocco, 41

Mosse, George L., 12
multinationality, 55, 121–22
Murgades i Barceló, Josep, 185

nationalism
Basques and, 74, 187
Catalunya and, 49–57, 81, 91
compared with racism, 40
democratic, 50–51
early theories of, 46
See also Catalan nationalism
Nations and Nationalism (Gelner), 37
Nietzsche, Friedrich, 46
Newtonian theory, 8
Northern Catalunya
associations in, 171
Catalan in, 144–47
as part of France, 133–34
nova cançó, 61, 161
Noyes, Dorothy, 127–28
Nueva Planta, 42, 45

Occitan, 49, 73, 85, 133, 144–45, 147, 152, 156, 165, 168
Olympic Games (1992), 49, 99
Omnium Cultural, 44, 171–72
Ortega y Gasset, José, 42

Pact of Tortosa, 46
Palau de la Música, 46, 58, 60
Palma, 3, 129–30
German population, 130
Partit Popular de Catalunya, 18, 123
Patum, 127–28, 129
penya, 29, 98, 107
See also football
Perpignan
Catalan and, 68, 74, 158, 160, 171–72
Cerbère and, 79–81
education and, 152–53, 165–66
gypsies and, 100, 134, 145–46
naming of children and, 140–42

Perpignan—*continued*
  Prades and, 125
  road improvement, 24
  rugby team, 30
  St. Jean festival and, 31–32
  TGV and, 23
  UCE and, 162–63
Picasso, Pablo, 137
Pieds Noirs, 66, 68
Pla, Josep, 65, 91, 116
Port de la Selva, 20, 103
Portbou
  associations in, 97, 99
  Cerbère and, 13–14, 25–35:
    symbolic links, 29–30;
    festival of St. Jean, 30–32;
    proposed tunnel, 24, 25, 91
  civic center, 91, 100, 103, 106, 107, 115
  commerce, 27–28
  demographics, 25–26
  German influence, 22, 34, 99, 102
  identity, 67
  overview, 20–25
  port construction, 24, 100, 101–2, 103, 106–7, 109, 111
  social life, 28
  sports, 28–29
  tourism, 27
postmodernism, 7–9, 10, 35
power, central vs. regional, 41–44
Prades, 120–21, 123–25, 134, 138, 152, 155
  UCE and, 163, 169, 172–73
Prats de Molló, 84, 134–35, 169
Primo de Rivera, Miguel, 47, 52, 63
Principat
  autonomy and, 48, 122, 126, 151, 158
  Catalanism and, 49, 57, 116–17, 120, 150
  Cerbère and, 79, 81, 91
  education and, 167, 168
  language and, 48–49, 76, 125, 165, 175, 187–88
  Portbou and, 115
  Sardana and, 58, 171

  St. Joan festival and, 32
  UCE and, 159, 163
  València and, 185
Pujol, Jordi, 5, 22–23, 47, 58, 124, 164, 181

Quintà, Alfons, 148

racism, 12–13, 40, 63–64, 189
rail workers, 34, 66–69, 71–72, 74–75, 85–90, 111
  See also transitors; transboarders
railroad
  Cerbère and, 15–16, 17, 23, 27, 69, 71
  mechanization of labor, 15, 66, 68, 76
  Northern Catalunya and, 151
  Portbou and, 20, 21–22, 23, 93, 95, 96, 100, 101, 103, 107, 116
  See also labor unions; rail workers; TGV
regional languages, 4, 29, 39, 137, 142, 144, 148, 166, 168, 186, 187
RENFE (Spanish rail system), 20, 24
Roca, Maria Mercè, 14, 93, 175
  interview, 176–78
Roig, Montserrat, 181
"Roots of the National Question in Spain" (Barton), 41
Rovira I Virgili, Antoni, 52

Sant Joan festival, 30–32, 60, 110, 149
Sant Jordi festival, 60, 153
Sardana, 28, 30, 58–59, 60, 74, 75, 77, 80, 84, 87, 90, 123, 171
Segadors, 57–58
Serrat, Joan Manuel, 61
Shakespeare, William, 2, 9
smuggling, 14, 71
SNCF (French National Railroad), 24, 66, 68, 80
soccer. *See* football

Socialist Party of Catalunya (PSC), 47, 122, 125, 156
Socialist Party of Spain (PSOE), 49, 183
Solidaritat Catalana, 52
Spanish American War, 42
Spanish Civil War, 46–47, 55, 58, 66, 71, 72, 73, 75, 88, 99, 104, 195
  refugees, 66, 71, 135–36, 160
St. Jean. *See Sant Joan* festival
St. Laurent de Cerdans, 135–36
Stegmann, Tilbert, 122, 175, 180–83
Strubell i Trueta, Miquel, 57

taxes, 16, 57–58, 67, 100, 151, 179
  added value, 16, 34
  revenues, 43–44
television, 18–19, 48, 69, 83, 96, 119, 155, 159, 165, 176, 182
temporality, 39
Terradas, Ignasi, 53–54
TGV (high speed rail line), 23, 69, 87
totalitarianism, 50, 110
Toubon, Jacques, 5
tourism, 13, 15–16, 20, 21, 23–24, 26–27, 34–35, 129–30, 134
  Cerbère and, 67, 69–70, 74, 79, 84, 85, 88
  Céret and, 137
  Northern Catalan and, 171
  Portbou and, 95, 99–100, 102, 103, 106, 107, 109, 111
  Prades and, 138
Trallero, Manuel, 126
transboarders, 71, 72, 74–76, 79
transborder cooperative project, 13
Transfesa, 95, 104
  effect on job market, 66, 68, 72, 76, 79
transitors, 13, 70–72, 74, 76, 78–79, 85, 87
Treaty of the Pyrénées, 133, 140
True France (Lebovics), 146
TV 3, 18, 48, 165, 176

Unamuno, Miguel, 43
underwater nature preserve, 15, 69, 86
United Front, 67
United Nations, 142, 144
  Catalan petition to, 191–96
United Nations Educational, Scientific, and Cultural Organization (UNESCO), 46, 60
Universitat Catalana d'Estiu (UCE)
  Catalanism and, 125–26, 158
  evolution of, 162–63
  formation of, 124, 155, 162
  La Razón article on, 125
  La Vanguardia article on, 125–26
  opinions of, 169, 181–82, 185
  overview of, 120–24
  Prades and, 163, 173
  search for Catalan culture in, 134–38
  summer session, 139

València, 18, 45, 61, 75, 126, 142, 159, 175, 181, 183–85, 187
Verdaguer, Jacint, 122
Viatge a les Entranyes de la Llengua (Casanova), 131, 144
video games, language and, 62
Vilaweb, 49
Villon, François, 156–57

Walter Benjamin Foundation, 91, 98, 111, 115
Walter Benjamin monument, 34, 98, 102, 106, 109
Weber, Max, 37
Wilson, Woodrow, 62
World War I, 8, 30, 75, 99, 159
World War II, 4, 15, 16, 66, 73, 85, 104, 136, 147, 160

Xarnegos, 112, 113

Yates, Alan, 122

Zapatero, José Luis Rodríguez, 148